Alternate Automotive Fuels

Alternate Automotive Fuels

Don Knowles
Saskatchewan Technical Institute

Reston Publishing Company, Inc.
A Prentice-Hall Company
Reston, Virginia

Library of Congress Cataloging in Publication Data

Knowles, Don.
　Alternate automotive fuels.

　Includes index.
　1. Synthetic fuels.　2. Motor fuels.　I. Title.
TP360.K58　1984　　　629.2′53　　　83-8680
ISBN 0-8359-0120-3
ISBN 0-8359-0119-X (pbk.)

Copyright © 1984 by
Reston Publishing Company, Inc.
A Prentice-Hall Company
Reston, Virginia 22090

All rights reserved. No part of this book may
be reproduced in any way, or by any means,
without written permission from the publisher.

10　9　8　7　6　5　4　3　2　1

Printed in the United States of America.

Contents

Preface — xiii
Abbreviations — xv

1 The Energy Crisis — 1

World Energy Reserves — 1
   Oil, Coal, and Gas Reserves — 1
U.S. Energy Reserves — 3
   Crude Oil Reserves — 3
   Natural Gas Reserves — 3
   Conventional Energy Resources and Consumption — 4
   Future U.S. Energy Requirements — 5
   Energy Cost Factors — 5
Canada's Energy Reserves — 7
   Crude Oil Reserves — 7
   Natural Gas Reserves — 7
   Canada's Future Energy Requirements — 7
Questions — 9

2 Propane—The Immediate Answer — 11

Propane Facts — 11
   Supply — 11
   Chemical Qualities — 13

Engine Design and Tuning Requirements ... 14
    Air Intake Temperature ... 14
    Air-Fuel Ratio Requirements ... 16
    Ignition Requirements ... 17
    Engine Design ... 17
    Engine Lubrication Requirements ... 17
    Cost Factors ... 18
Questions ... 19

3 Types of Propane Conversion Equipment — 21

Impco Propane Equipment ... 21
    Vaporizer Design ... 21
    Vaporizer Operation ... 21
    Priming Solenoids ... 23
    EC1-Equipped Vaporizers ... 23
    Mixer Design ... 26
    Mixer Operation ... 26
    Types of Mixers ... 26
    Mixer Selection ... 31
    Vacuum Fuelock ... 32
Garretson Propane Equipment ... 35
    Vaporizer Design ... 35
    Mixer Design ... 37
    Mixer Adjustment ... 38
    Idle Control Valve ... 38
Tartarini Propane Equipment ... 40
    Vaporizer Design ... 40
    Vaporizer Operation ... 41
    Mixer Design ... 44
    Mixer Operation ... 44
Century Propane Equipment ... 46
    Vaporizer Design ... 46
    Vaporizer Operation ... 46
    Model 2610 Mixer ... 48
Vialle Propane Equipment ... 51
    Vaporizer Design ... 51
    Vaporizer Operation ... 52
    Vaporizer Purge System ... 55
    Mixers ... 57

Petrosystems Propane Equipment	60
Mixers	60
Switching Valve	60
Converters	63
Propane Diesel Boosting Equipment	63
McCoy Mileage Master	63
Questions	64

4 Propane Conversion Installations — 67

Straight Propane Conversions	67
Motor Fuel Tanks	67
Liquid Fuel Lines	68
Vaporizer Installation	73
Mixer Installation	73
Dual Fuel Propane Conversions	75
Dual Fuel and Straight Propane System Variations	75
Fuelocks	77
Dual Fuel Wiring	80
Generally Accepted Regulations	83
Motor Fuel Tanks	83
Hose Requirements	87
Safety Precautions	88
Questions	89

5 Engine Tuning for Propane Fuel — 91

Ignition Basics	91
Ignition Operation	91
Secondary Voltage Requirements	94
Propane Ignition Requirements	94
Secondary Ignition Circuit	94
Distributor Timing and Advance	100
Purpose of Distributor Advance	100
Centrifugal Advance Operation	101
Vacuum Advance Operation	102
Propane Distributor Advance Requirements	102
Initial Timing	102
Distributor Advance Specifications	102
Effects of Incorrect Ignition Spark Advance	104
Calibrating Distributor Advances for Propane-Fueled Engines	106

Infrared Adjustment and Diagnosis of Propane Fuel Systems	107
Emission Levels	107
Adjusting Impco EC1-Equipped Vaporizers	110
Diagnosis of Propane-Fueled Engines	111
Hard Starting	111
Rough Idle Operation	111
Excessive Power Loss	113
Hesitation on Acceleration	113
Misfiring on Acceleration	113
Low Fuel Economy	113
Detonation and Pinging	114
Questions	114

6 Propane Conversions and Automotive Computer Systems — 115

A Computer Command Control (3C) System	115
Purpose	115
Computer Input Signals	115
Throttle Position Sensor (TPS)	116
Coolant Temperature Sensor	117
Pressure Sensors	117
Vehicle Speed Sensor (VSS)	120
Computer Output Signals	120
Carburetor Management	120
Ignition Management	121
Torque Converter Clutch (TCC) Management	121
Early Fuel Evaporation (EFE) Management	125
Air Injection Reactor (AIR) Management	127
Exhaust Gas Recirculation (EGR) Management	128
Idle Speed Management	129
Self-Diagnostics	129
Precautions for Propane Conversions on Computer Systems	132
Questions	134

7 Natural Gas—The Answer to Inexpensive Fuel — 137

Natural Gas Facts	137
Availability	137
Availability in the United States	139
Availability in Canada	141

Cost Factors	141
Composition and Tuning Requirements	143
Ignition Requirements	143
Carbon Monoxide Levels	146
Emission Levels	147
U.S. Experience with CNG	147
Fleet Experience	147
Italian CNG Experience	149
Generalizations	149
Approval of CNG Conversion Equipment	150
Approval of CNG Cylinders	150
Vehicle Conversion and Testing	151
Fueling Stations	151
CNG Safety Record	152
General	152
ANCC Safety Records	153
SNAM Safety Records	153
Conclusions	153
Questions	154

8 Compressed Natural Gas Conversion Equipment — 155

CNG Components	155
Fuel Cylinders	155
Lines and Fill Fitting	155
Pressure Regulators	155
Mixer	156
Tartarini CNG Equipment	159
Pressure Regulators	159
Mixers	160
Complete CNG Installations	161
Questions	163

9 Compressed Natural Gas—Regulations and Safety Precautions — 165

Generally Accepted Regulations	165
CNG Regulations	165
CNG Safety Precautions	169
Questions	170

10 Filling Facilities for Compressed Natural Gas — 171

Compressors	171
Reciprocating Compressors	171
Hydraulic Compressors	172
CNG Measuring Devices	175
U-Tube Magnetic Sensing Meters	175
Questions	175

11 Liquid Natural Gas — 177

General Facts	177
Natural Gas Liquefaction Procedure	177
BTU Content and Cost Factors of LNG	178
LNG Fuel Tanks	178
LNG Conversion Systems	180
Liquid Mode Operation	180
Vapor Mode Operation	182
LNG Electrical System	183
Engine Tuning Requirements	185
LNG Refueling Facilities	185
Pressure Transfer Filling Stations	185
In-Line Pump Filling Station	185
Submerged Pump Filling Stations	185
Questions	188

12 Alcohols as Motor Fuels — 189

Ethanol	189
Chemical Composition and BTU Content	189
Feedstock	189
Processing Plants and Cost Factors	190
Gasohol	192
Methanol	194
Chemical Composition and BTU Content	194
Feedstock	195
Processing Plants and Cost Factors	195
The Brazilian Experience	195
Ethanol Production	195
Ethanol Automotive Conversions	196

Advantages of Alcohols	196
Disadvantages of Alcohols	197
Questions	198

13 Engine Tuning, Fuel Economy, and Exhaust Emissions with Alcohol Fuels — 199

Fuel System and Ignition Tuning	199
Air-Fuel Ratio Requirements	199
Carburetor Modifications	199
Ignition Tuning	201
Alcohols in Diesel Engines	202
Fuel Mileage and Emission Levels with Alcohols	203
Fuel Economy	203
Evaporative Emissions	203
Nitrous Oxide Emissions	203
Carbon Monoxide Emissions	204
Hydrocarbon Emissions	205
Aldehyde Emissions	205
Savannah River Plant Alcohol Program	206
Fleet Size	206
Gasohol Blends	206
Problems Encountered	207
Cost Factors	207
Conclusions	207
Engine Wear and Methanol Fuel	208
U.S. Army Test Results	208
Questions	208

14 Hydrogen—The Fuel of the Twenty-First Century — 209

Hydrogen Development	209
General Facts	209
Sources of Hydrogen	209
Cost Factors	210
Hydrogen from Solar Energy	210
Hydrogen as an Aircraft Fuel	210
Hydrogen as an Automotive Fuel	210
Fuel Storage	210
Liquid Hydrogen Project	211
Hydrogen Injection Project	215

	SLX Hydrogen Breakthrough	217
	Exhaust Emissions with Hydrogen Fuel	218
	Safety and Hydrogen Fuel	218
	Questions	220

15 Future Energy Sources — 223

Solar Energy	223
Potential of Solar Power	223
Wind Energy	223
Development	223
Technology from the Sea	224
The Fusion Solution	224
Energy from the Sea	227
Ocean Thermal Energy Conversion	227
Questions	227
Index	229

Preface

Gasoline shortages and dramatic price increases in the 1970s created a growing interest in fuels that could replace gasoline. The depressed economy in recent months has influenced many companies to reduce expenses in every way possible. Many companies involved in the transportation sector have experienced substantial savings by converting their vehicles to alternate fuels. Each year we find an increasing number of vehicles in the United States and Canada powered by some type of alternate fuel.

The purpose of this book is to explain the various alternate fuel systems available at the present time and to provide the reader with the advantages, disadvantages, cost factors, safety precautions, and applicable regulations connected with each fuel. The alternate fuel systems included in the book are propane, compressed natural gas (CNG), liquid natural gas (LNG), alcohols, and hydrogen.

Energy reserves in the United States and Canada will determine, to a large extent, the type of alternate fuels that will be developed. This book provides an outline of energy reserves in the United States and Canada in relation to world energy reserves.

Many people are concerned about the possibility of engine damage from burning alternate fuels. Engine timing specifications, distributor advance requirements, and air-fuel ratio adjustments for the different alternate fuels are provided in detail. Costly mistakes could be avoided by following the engine tuning information in the book.

We can derive much benefit from the experiences of other nations with alternate fuels. The Italian experience with CNG and the Brazilian alcohol program are examples of alternate fuel developments in other nations that are provided in the text.

I would like to thank all the people in the alternate fuel industries for their assistance in supplying diagrams and information for the book. A special thank you must go to everyone at Reston Publishing for their excellent cooperation in producing the finished work.

Abbreviations

AIR	air injection reactor
ATDC	after top dead center
bbl	barrels
BTDC	before top dead center
BTU	British thermal unit
cfm	cubic feet per minute
CID	cubic inch displacement
CNG	compressed natural gas
EFE	early fuel evaporation
EGO	exhaust gas oxygen
EGR	exhaust gas recirculation
EST	electronic spark timing
GH	hydrogen gas
gpm	grams per mile
HC	hydrocarbon
hp	horsepower
ICE	internal combustion engine
kj	kilojoule
kp	kilopascal
kv	kilovolt
l	liter
LH	liquid hydrogen
LNG	liquid natural gas
LPG	liquid propane gas

Abbreviations

MAP	manifold absolute pressure
NOx	nitrous oxide
OH	hydroxy radical
OTEC	ocean thermal energy conversion
ppm	parts per million
psi	pounds per square inch
rpm	revolutions per minute
TCC	torque converter clutch
TDC	top dead center
TPS	throttle position sensor
VPS	vapor purge system
VSS	vehicle speed sensor

Alternate Automotive Fuels

CHAPTER 1

The Energy Crisis

World Energy Reserves

OIL, COAL, AND GAS RESERVES The energy reserves of a nation will determine to a large extent the type of motor fuel each nation will use in the future. Another factor in the selection of future fuels will be the economics of obtaining different fuels from the energy reserves. Many nations will not become energy self-sufficient unless they develop alternate fuels and unconventional petroleum sources.

Proven recoverable world energy reserves of oil, coal, natural gas, and uranium are pictured in Figure 1-1. The U.S. share of the world reserves is also illustrated. One quad is equal to one quadrillion BTUs of energy. Crude oil reserve figures may vary widely. Some statistics are based on proven easily recoverable oil, while other figures quote the supply of oil that is actually in a given country or oil field. The recovery rate varies from 5 to 60 percent in different oil fields. Undeveloped oil fields may be included in some oil reserve statistics.

Coal reserves far exceed other energy sources. The United States has about 25 percent of the world's coal reserves. World consumption of crude oil is approximately 134 quadrillion BTUs per year, and world crude oil reserves are 3,721 quadrillion BTUs; thus, a thirty-year supply of crude oil exists at present. Some geologists estimate that through exploration they will discover new oil deposits to maintain present reserves. Other experts believe oil consumption will continue to rise and reserves will be depleted sooner than expected.

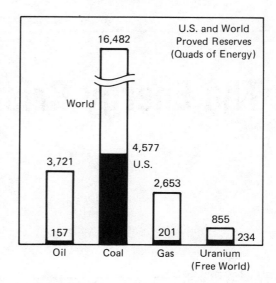

FIGURE 1-1. World Energy Reserves.

TABLE 1-1. World Crude Oil Reserves

	Billions of bbl	Billions of m^3
1. Middle East	361.8	57.4
2. Sino-Soviet	78.0	12.3
3. Africa	57.1	9.0
4. Mexico and Latin America	56.5	8.9
5. United States and Canada	33.3	5.2
6. Western Europe (North Sea)	23.6	3.7
7. Asia	17.1	2.7
8. Oceana	2.2	0.3

The location of proven recoverable world crude oil reserves is listed in Table 1-1. Middle East crude oil reserves are much greater than the reserves in other parts of the world.

U.S. Energy Reserves

CRUDE OIL RESERVES Crude oil reserves from conventional and unconventional sources are listed in Table 1-2. One significant factor is the huge shale oil reserves in the United States. Shale oil reserves are largely undeveloped at the present time.

TABLE 1-2. U.S. Crude Oil Reserves

	Billions of bbl	Billions of m^3
1. Proven	27	4.2
2. Potential	133	21.0
3. Shale oil	1,026	162.0
4. Tar sands	15	2.3
5. Heavy oil	30	4.7

The United States consumes approximately 13 million bbl of crude oil per day. Domestic production is about 8.5 million bbl, and the remaining crude oil required is imported. Crude oil consumption has declined in recent years mainly because smaller, more fuel-efficient cars are now widely available. U.S. oil imports were down 23 percent in a recent eight-month period, and domestic exploration and production increased significantly.

NATURAL GAS RESERVES Conventional and unconventional U.S. natural gas reserves are listed in Table 1-3.

Huge natural gas reserves exist in geopressured zones in the United States. Much of the natural gas from this source is located at 20,000–30,000 ft (6,096–9,144 m). Exploration and development will therefore be expensive compared with that for conventional sources.

TABLE 1-3. U.S. Natural Gas Reserves

	Trillions of ft^3	Trillions of m^3
Conventional	195	5.6
Potential	1,019	29.0
Geopressured zones	3,000	85.6
Tight sands	800	22.8
Devonian shale	600	17.0
Coal seams	500	14.0

CONVENTIONAL ENERGY RESOURCES AND CONSUMPTION U.S. energy reserves of oil, gas, and coal are listed in Table 1-4.

TABLE 1-4. U.S. Energy Reserves

	Quads
1. Coal	4577
2. Gas	201
3. Crude oil	157

The approximate yearly consumption of oil, gas, and coal in the United States is provided in Table 1-5.

TABLE 1-5. U.S. Energy Consumption

	Quads	Percentage of Conventional Reserves
1. Gas	20.0	10
2. Oil	18.0	11
3. Coal	17.5	0.4

The supply of proven easily recoverable oil and gas is diminishing rapidly. Extensive new discoveries, or development of unconventional sources are necessary to meet future energy requirements.

FUTURE U.S. ENERGY REQUIREMENTS Some interesting developments will take place on the energy scene during the next two decades. There will be a gradual shift from the use of petroleum-derived energy to the use of other energy sources, particularly solar energy. Dependence on imported crude oil will decline and the use of coal will increase. Table 1-6 lists the estimated U.S. energy requirements through the year 2000.

TABLE 1-6. Future U.S. Energy Requirements

	1980 (quads)	1990 (quads)	2000 (quads)
Domestic crude oil	20.5	19.6	20.5
Imported crude oil	15.2	11.7	13.1
Gas	20.5	19.4	17
Nuclear	2.9	8.1	11.3
Other	3.1	3.7	11.7
Coal	15.8	26.5	34.4
TOTALS	67	78	108

ENERGY COST FACTORS Many questions arise regarding future energy developments. What type of energy will be developed from the huge coal reserves? Which alternate fuel will be the most economical to produce? Some of the future sources of fuel are illustrated in Figure 1-2. Notice that hydrogen can be made by gasification of coal, or electrolysis of water. Methanol, gasoline, or methane (natural gas) can be obtained by coal gasification. Gasoline can also be derived from methanol. All the fuels in Figure 1-2 can be used to power the internal combustion engine (ICE).

The types of fuels used in the future will depend largely on the supply of natural resources, and on the cost of producing various fuels. Environmental impacts of pollutants arising from the production of different fuels and tail pipe emissions will be a factor in the choice of alternate fuels. Pollution from the increased use of coal is another concern to environmentalists. Cost factors in the production of some alternate fuels are outlined in Figure 1-3. Refining gasoline from petroleum involves an energy conversion process that is 89 percent efficient and costs $5 per million BTUs. Producing hydrogen by electrolysis has an energy conversion efficiency of 23 percent and costs $25 per million BTUs. Producing natural gas or methanol by coal gasification would appear to be the least costly alternate source of energy. With the present fast-changing technology, energy production

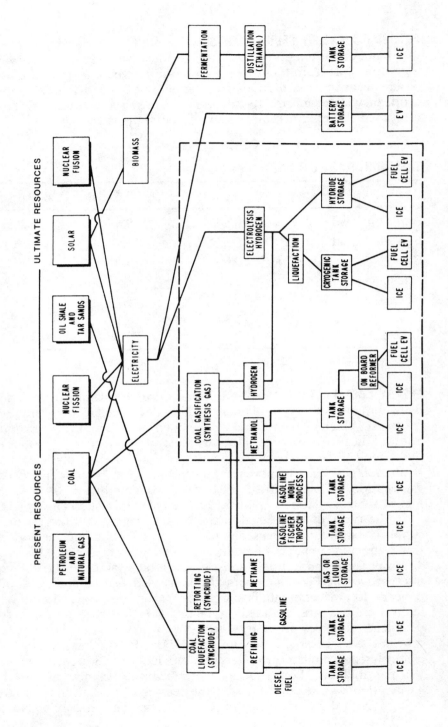

FIGURE 1-2. Energy Sources. *(Reprinted with permission © 1981, Society of Automotive Engineers)*

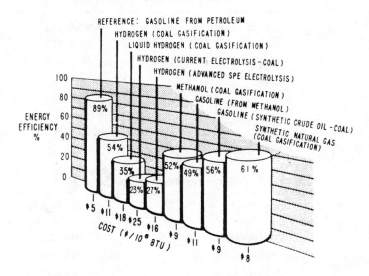

FIGURE 1-3. Energy Cost Factors. *(Reprinted with permission © 1981, Society of Automotive Engineers)*

costs may also change rapidly. Conversion of a significant percentage of the U.S. car and light truck fleet to natural gas or methanol could play a vital role in helping the world to become energy self-sufficient.

Canada's Energy Reserves

CRUDE OIL RESERVES Current reserves, production, and consumption of crude oil are listed in Table 1-7.

NATURAL GAS RESERVES Canada has large reserves of natural gas, as indicated in Table 1-8.

CANADA'S FUTURE ENERGY REQUIREMENTS The Geological Survey of Canada estimates that 2.52 billion bbl (0.4 billion m^3) of crude oil will be discovered in western Canada in the 1980s. Natural gas discoveries will total

TABLE 1-7. Crude Oil Statistics for Canada

	Millions of bbl	Millions of m^3
Initial reserves in place	14,408	2,287
Remaining	5,077	806
Yearly production	648	103
Consumption	686	109
Imports	220	35
Exports	176	28

TABLE 1-8. Natural Gas Statistics for Canada

	Billions of ft^3	Billions of m^3
Initial reserves in place	104,405	2,983
Remaining	68,460	1,956
Production	2,443	69.8
Consumption	1,666	47.6
Exports	791	22.6
Imports	0	0.0

35 trillion ft^3 (1 trillion m^3) in the same period. Production of conventional crude oil is expected to decline from 1,493,730 bbl (237,100 m^3) per day in 1980 to 1,173,690 bbl (186,300 m^3) per day in 1985. The Athabasca Tar Sands in western Canada are thought to contain the world's largest hydrocarbon deposits, which are estimated at 201 billion bbl (32 billion m^3) of recoverable crude oil. Two syncrude plants are operating in the tar sands at present. Arctic and east coast offshore reserves are estimated at 66.5 trillion ft^3 (1.9 trillion m^3) of natural gas and 5,203 million bbl (826 million m^3 of crude oil). With conventional oil reserves in western Canada dwindling rapidly, the development of Arctic, offshore, and tar sands reserves is an important means of achieving energy self-sufficiency. Conversion of a large number of vehicles to propane, compressed natural gas, or liquid natural gas could also play a significant role in this effort. Canada's estimated energy requirements for this decade are outlined in Table 1-9. Predictions indicate a decline in the use of crude oil and an increase in the use of natural gas and renewable energy.

TABLE 1-9. Canada's Future Energy Requirements

	Percentages	
Energy Source	1980	1990
Oil	42	27
Gas	18	23
Electricity	27	32
Coal	9	10
Renewables	3	6
LPGs	1	2
Total energy in petajoules	9,525	11,900

Questions

1. The largest U.S. energy reserves are in _____.
2. At current consumption, the world supply of proven recoverable crude oil will last approximately _____ years.
3. The United States has more shale oil reserves than any other unconventional source.   T   F
4. Natural gas reserves in geopressured zones exceed all other sources in the United States.   T   F
5. Total energy consumption in the United States will decline in the next two decades.   T   F
6. Methanol can be made by coal gasification.   T   F
7. The largest hydrocarbon deposit in the world is the _____.
8. Producing methanol by coal gasification is less expensive than producing hydrogen by current electrolysis.   T   F

CHAPTER 2

Propane—The Immediate Answer

Propane Facts

SUPPLY Sixty-five percent of the supply of propane in the United States comes from natural gas wells and natural gas processing plants. Propane from this source is often referred to as wet gas or wellhead gas. The remaining 35 percent of the U.S. propane supply is obtained in the gasoline refining process. Proven reserves of gas liquids, propane, butane, and so on declined in the United States during the 1970s, as indicated in Figure 2-1. An increasing demand for propane is predicted for the 1980s, while the domestic supply will continue to decline, as illustrated in Figure 2-2. If large numbers of vehicles were converted to propane in the United States, the dependence on imported propane might increase along with prices.

Canada produces approximately 120,000 bbl (19,047 m^3) of propane per day, and consumption has been about 50,000 bbl (7,936 m^3) per day. Surplus propane has been exported. Producers have sometimes wasted propane by burning it off, a process that is commonly referred to as flaring. Saudi Arabia has flared off as much as 4 billion ft^3 (114,285,714 m^3) of natural gas per day. No doubt this gas contained significant quantities of propane. A vast collection system now saves these gases for export.

In most parts of the United States and Canada, propane provides an alternate fuel that is available immediately. A propane distribution network already exists in these countries to supply home heating, industrial, and recreation vehicle (RV) requirements. Some propane

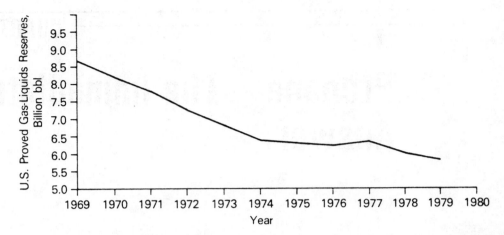

FIGURE 2-1. U.S. Gas Liquid Reserves. *(Reprinted with permission © 1982, Society of Automotive Engineers)*

outlets, however, would have to be upgraded with meters and hoses to fill motor fuel tanks. Propane conversion equipment is readily available to take care of improvements in distribution networks or filling facilities that might be necessary in certain areas.

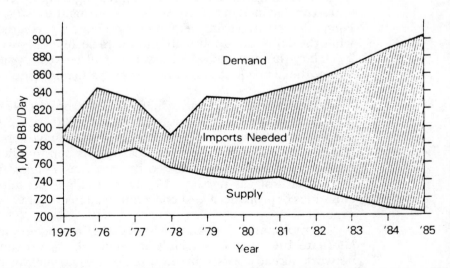

FIGURE 2-2. U.S. Propane Supply and Demand. *(Reprinted with permission © 1982, Society of Automotive Engineers)*

Some governments offer financial incentives to encourage propane conversions. The Canadian government, for example, offers a $400 rebate to fleet operators and farmers for each straight propane conversion. Program objectives are 100,000 propane conversions by 1985. In 1981 propane outlets in Canada increased from 350 to 935. At the same time, factory-equipped vehicles have become available in both the United States and Canada. Ford Motor Co. was the first major car manufacturer to offer propane-powered vehicles; Granadas and Cougars are available with factory-installed propane equipment.

CHEMICAL QUALITIES Propane is a hydrocarbon fuel that differs in molecular structure from gasoline. The chemical symbol for propane is C_3H_8. Each molecule of propane contains three carbon atoms and eight hydrogen atoms. The molecular composition of gasoline is C_8H_{15}. Thus, propane has a lower carbon-to-hydrogen ratio. Hydrocarbon fuels with a lower carbon content generally have a higher octane rating and less tendency to detonate in the engine. As a result, propane fuel is known for its cleaner combustion and reduced oil contamination. The weight of propane is approximately 4.25 lb (1.9 kg) per U.S. gal, or 5.08 lb (2.2 kg) per imperial gal. Gasoline weighs about 6 lb (2.7 kg) per U.S. gal, or 7.2 lb (3.2 kg) per imperial gal.

Propane contains 109,800 BTUs (111,439 kj) per imperial gal, or 91,500 BTUs (96,532 kj) per U.S. gal. The BTU content of gasoline varies from 113,000 to 147,000 (119,000–155,000 kj), depending on the grade of gasoline and whether U.S. or imperial gallons are quoted. Some engines converted to propane may experience a 5–10 percent power loss. If engines were optimized for propane fuel, the power loss would be eliminated.

The vapor pressure of propane varies in relation to atmospheric temperature:

```
100°F   (38°C)  —  190 psi  (1330 kp)
 70°F   (21°C)  —  120 psi   (840 kp)
  0°F  (−18°C)  —   28 psi   (196 kp)
−20°F  (−30°C)  —   13 psi    (91 kp)
−40°F  (−40°C)  —    2 psi    (14 kp)
−45°F  (−43°C)  —    0 psi     (0 kp)
```

The boiling points of liquid propane (LPG), compressed natural gas (CNG), gasoline, and diesel fuel are compared in Figure 2-3. Propane will boil or vaporize at −44°F (−42°C), while CNG boils at −260°F (−162°C).

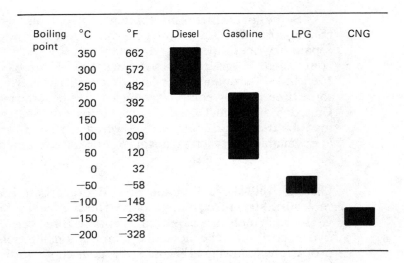

FIGURE 2-3. Hydrocarbon Fuel Boiling Points.

At temperatures below −45°F (−43°C), pressure in the propane fuel tank will be zero. Tank pressure is used to force the propane from the fuel tank to the vaporizer. The fuel system will not operate at temperatures below −45°F (−43°C). This is not a problem in most climates. During hot weather fuel tank pressure could reach 190 psi (1330 kp). In most countries externally mounted propane tanks are rated at 250 psi (1750 kp), and internally mounted tanks have a rating of 312 psi (2184 kp). During the manufacturing process, the tanks are tested at a much higher pressure. Propane motor fuel tanks have excellent safety margins.

Engine Design and Tuning Requirements

AIR INTAKE TEMPERATURE Propane fuel is converted from a liquid to a vapor in the vaporizer and delivered to the mixer and engine cylinders in gaseous form. Engines fueled with propane will operate more efficiently with lower air intake temperatures. Propane vapor becomes more dense with cooler air intake temperatures. Propane conversion equipment should include an enclosed air cleaner with an intake tube extending behind the vehicle grill to provide a cooler supply of intake air, as illustrated in Figure 2-4.

FIGURE 2-4. Enclosed Air Cleaner.

An open-style air cleaner takes in extremely hot air under the hood. For every 10°F (5.5°C) rise in air intake temperature above atmospheric temperature, a 1 percent power loss will be experienced. Engine power will be decreased 10 percent if a 100°F (37°C) difference exists between air intake temperature and atmospheric temperature. High intake air temperature contributes to detonation in the combustion chambers. Cooler air intake temperatures allow additional spark advance and improve economy and performance.

Blocking the heat riser valve in the open position or plugging the intake manifold exhaust crossover passage will reduce exhaust flow through the crossover passage and lower the intake manifold temperature. Factory-equipped propane-fueled vehicles have intake manifolds designed to operate at lower temperatures. When propane fuel is used, there is no concern about cold intake manifolds causing fuel condensation since propane vapor can only be condensed by pressurizing the vapor.

Proper operation of the cooling system is also important in a propane-fueled engine. Engine thermostats rated above 180°F (82°C) should never be used. If the temperature inside the vehicle is not adversely affected, a 165°F (74°C) thermostat would be preferred. It has been estimated that 1 in (2.5 cm) of cast iron coated with 1/16 in (1.5 mm)

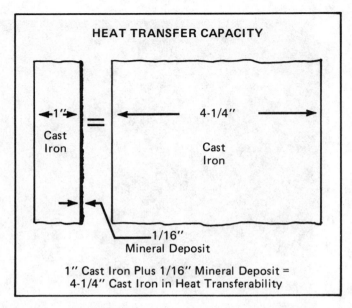

FIGURE 2-5. Effects of Cooling System Deposits. *(Courtesy of Impco Carburetion, Inc.)*

of mineral deposits has the same heat transfer capacity as 4¼ in (10.6 cm) of cast iron, as indicated in Figure 2-5.

AIR-FUEL RATIO REQUIREMENTS Correct air-fuel ratios are critical in a propane-fueled engine. Propane vapor has very little cooling effect on engine parts. A rich air-fuel mixture can cause engine overheating, as shown in Table 2-1.

TABLE 2-1. Effects of Rich Mixture On Engine Temperature.

	RPM	Load, BMEP	Exhaust* Temp°F	Valve Face°F	Valve Head°F
Mixture: (Air-Fuel ratio)					
Rich	1200	165	1342	1338	1378
Lean**	1200	165	1288	1285	1336
			54	53	42

SOURCE: Courtesy Impco Carburetion Inc.

Since lean propane air-fuel ratios result in slower combustion in the cylinders, excessive burning continues to take place when the exhaust valve opens. As a result, exhaust valve temperature is increased and valve life is shortened. Correct air-fuel ratio adjustments are discussed in Chapter 5.

IGNITION REQUIREMENTS A propane-fueled engine must have correct distributor advance curves and initial timing settings. Excessive distributor advance results in detonation and engine damage. Engine overheating, power loss, and reduced engine life can be caused by insufficient spark advance. Propane fuel requires additional spark advance at low rpm. (Chapter 5 provides actual figures on timing and distributor advance requirements.)

ENGINE DESIGN Engine compression ratios could be increased to gain optimum power and economy from propane fuel. Compression ratios on gasoline engines are now approximately 8.2:1. The introduction of unleaded gasoline and catalytic converters forced engine manufacturers to lower the compression ratios on gasoline engines. Propane has an octane rating of 112 compared with gasoline's octane rating of 90–92. Engines designed for propane fuel could have the compression ratio increased to 10.5:1 without detonation problems. Some manufacturers of factory-equipped propane-fueled vehicles have increased the compression ratios. Propane is a cleaner burning fuel than gasoline, and in most cases exhaust emission requirements can still be met with the higher compression ratios. A propane-fueled engine does not require a choke for cold mixture enrichment because the propane enters the intake manifold in a gaseous state. Sudden acceleration no longer requires a richer mixture, and the accelerator pump is eliminated in the propane mixer. Normal propane air-fuel mixtures are leaner than gasoline mixtures. A propane-fueled engine will operate more efficiently than a gasoline-fueled engine. The gain in efficiency is most noticeable at low rpm.

ENGINE LUBRICATION REQUIREMENTS The average high-detergent, multiple-viscosity engine oil contains barium and calcium additives. Gum and sulphur deposits from gasoline are suspended in the oil by the barium and calcium additives. Propane fuel, on the other hand, does not contain gum or sulphur deposits, so that the oil additives have nothing to combine with in a propane-fueled engine. Under certain conditions, the oil additives can burn in the combustion chambers and thus create deposits on valves and spark plugs. This problem is more likely to occur in a heavy duty application, where more cylinder heat is encountered. Incorrect air-fuel mixtures or distributor advance curves create excessive combustion chamber heat and promote oil additive burning. Many oil companies produce a specially formulated low additive oil for propane-fueled engines.

Some propane equipment manufacturers do not recommend the use of an ashless oil (synthetic oils are classified as ashless oils) with propane fuel. Ashless oil can contribute to valve seat recession because the reduced carbon content of propane fuel provides less cushioning action between the valve face and seat. Small particles of the valve seat can become embedded on the valve face. The process occurs each time the valve opens, so that eventually the valve seat deteriorates. Valve seat recession is not a problem when propane fuel is used in engines designed for unleaded gasoline since these engines have harder valve seats.

COST FACTORS The price difference between propane and gasoline varies widely within a country. In some locations in North America propane sells for half the price of gasoline, while in other areas both fuels are the same price. Propane costs exceed gasoline prices in a few areas. The cost of vehicle conversion also varies, depending on the cost of labor, cost of conversion equipment, and the size of tank installed. Remote-fill hoses will also raise the cost of conversion. Passenger car conversions might cost about $1,000 to $1,800. Small truck conversions usually cost about $200 less than car conversions because remote-fill hoses are not required. The average cost of a passenger car conversion in the United States would be $1,000. Let us assume that the price spread between propane and gasoline is 45¢ per gal (10¢ per l). If a vehicle averaged 20 mpg, 50 gal (225 l) of fuel would be consumed in 1,000 mi (1,600 km). A fuel saving of $22.50 would be realized every 1,000 mi (1,600 km). If the vehicle was operated for 40,000 mi (64,000 km), the fuel saving would be $900. Motor homes or trucks averaging 10 mi per gal would provide an equivalent fuel saving in 20,000 mi (32,000 km). Fuel savings will have to be calculated for each area and will depend on price differentials and conversion costs. In many areas, payback periods for propane conversions could be about 20,000-40,000 mi (32,000-64,000 km). Other savings such as those in the form of fewer oil and filter changes and longer engine life are not considered in the above calculations. Propane conversion equipment will outlast the vehicle on which it is installed. When vehicles are replaced, the conversion equipment can be installed on the new vehicles. This fact is not taken into account in our cost-saving figures.

Questions

1. In the past decade the U.S. domestic propane supply has increased each year.   T   F
2. A molecule of propane contains _____ carbon atoms and _____ hydrogen atoms.
3. Propane has a lower carbon-to-hydrogen ratio than gasoline.   T   F
4. The boiling point of propane is _____ degrees F.
5. The pressure in a propane fuel tank would be _____ psi if the atmospheric temperature is 100°F (38°C).
6. Compressed natural gas has a higher boiling point than propane.   T   F
7. Propane-fueled engines operate more efficiently with cold air intake temperatures.   T   F
8. Propane has an octane rating of _____.
9. A propane-fueled engine requires the same ignition spark advance as an engine operating on gasoline.   T   F
10. An engine oil with barium and calcium additives is required in a propane-fueled engine.   T   F

CHAPTER 3

Types of Propane Conversion Equipment

Impco Propane Equipment

VAPORIZER DESIGN When the fuelock is energized, pressurized liquid propane flows through the propane fuelock and filter to the vaporizer. The vaporizer reduces the pressure of the liquid propane, changes the liquid propane to a vapor, and delivers the gaseous fuel to the mixer. Most vaporizers are two-stage units that contain a primary section and a secondary section. As illustrated in Figure 3-1, the secondary diaphragm area is connected to the mixer.

The manual primer is connected to the cover side of the secondary diaphragm. Some vaporizers use electric primers, or a combination of manual and electric primers. Heater hoses from the cooling system circulate coolant through the water passages in the vaporizer. Expansion chambers prevent damage to the vaporizer if the coolant freezes.

VAPORIZER OPERATION When propane flows through the open primary seat, it changes to a vapor because its pressure is reduced. Heat from the cooling system assists in converting the liquid propane to a vapor. When the primary pressure reaches 3-5 psi (21-35 kp), the primary diaphragm is pushed over against the primary spring tension. The primary

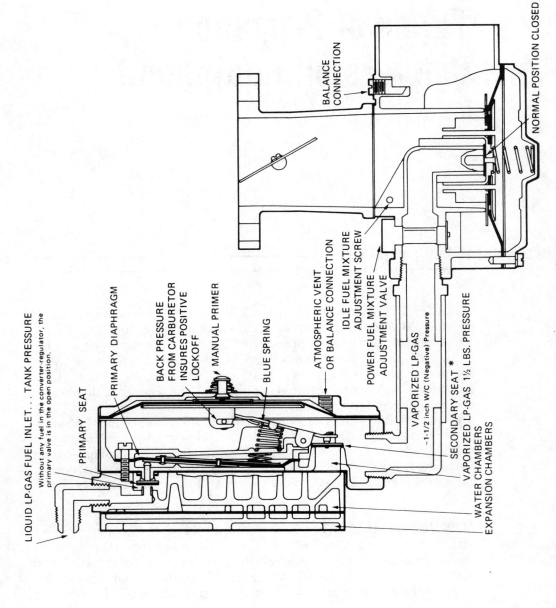

FIGURE 3-1. Propane Vaporizer. *(Courtesy of Impco Carburetion, Inc.)*

diaphragm moves on its pivot and closes the primary valve, thus preventing further flow of propane into the vaporizer.

Because of the large area of the secondary diaphragm and the mechanical advantage of the secondary lever, a slight vacuum from the mixer can open the secondary valve. When the secondary valve opens, propane vapor can move into the mixer and engine cylinders. The vacuum in the mixer and secondary section increases in relation to the engine rpm. The increased vacuum gradually pulls the secondary valve open as engine speed increases. When propane vapor is moved through the secondary valve, the primary pressure decreases. A decrease in primary pressure allows the primary spring to move the primary diaphragm and open the primary valve so that propane can enter the primary section. When the engine is stopped, the secondary valve closes, and thus prevents any further flow of propane vapor. The horsepower rating of the vaporizer must be matched to the engine horsepower. Many vaporizers are rated for engines up to 350 hp.

PRIMING SOLENOIDS Most vaporizers use a prime solenoid to facilitate cold starting. The prime solenoid (component A in Figure 3-2) is threaded into the secondary section of the vaporizer.

A vacuum hose connects the solenoid to the intake manifold below the throttle plates. One solenoid terminal is connected to ground. The other solenoid terminal is connected to the main terminal of the starting motor so that the solenoid will operate each time the engine is started. The prime solenoid terminal may be energized by a dash-mounted prime button. Once the prime solenoid is energized, the manifold vacuum can move propane vapor directly into the intake manifold from the secondary section. The flow of propane vapor through the priming system facilitates cold engine starting. The small solenoid vacuum hose will fill with propane vapor much faster than the large vapor hose from the vaporizer to the mixer.

EC1-EQUIPPED VAPORIZERS Impco vaporizers may be obtained with an EC1-equipped secondary diaphragm cover. The EC1 device may be retrofitted to most Impco vaporizers. Leaner air-fuel ratios may be obtained with the EC1-equipped vaporizer. Without the EC1 device, an atmospheric vent in the secondary diaphragm cover allows atmospheric pressure to be applied between the secondary and cover. Movement of the secondary diaphragm is controlled by the mixer vacuum on one side and atmospheric pressure on the other side. The EC1 equipment controls the atmospheric pressure between the secondary diaphragm and the cover during cruise conditions. A lowering of the atmospheric pressure

FIGURE 3-2. Priming Solenoid. *(Courtesy of Emco Wheaton, Ltd.)*

will reduce secondary diaphragm and valve movement, and this reduction will result in leaner air-fuel ratios. An EC1-equipped vaporizer cover is illustrated in Figure 3-3. The EC1 device may be used to meet emission regulations or improve fuel economy.

A schematic diagram of the EC1 is presented in Figure 3-4.

Bleed screw A controls the amount of atmospheric pressure applied to the secondary diaphragm. The mixer vacuum is connected to port B. If screw A is turned outward, atmospheric pressure on the secondary diaphragm will increase. The increase in secondary diaphragm movement provides additional secondary valve opening and richer air-fuel ratios. Atmospheric pressure on the secondary diaphragm is reduced when screw A is turned inward. The reduced secondary valve opening and leaner air-fuel ratios result from the reduced atmospheric pressure on the diaphragm. A high manifold vacuum during idle or cruise conditions holds diaphragm D and lever E in the upward position. When diaphragm D is held upward, lever E will not touch valve F. Under full load conditions, a reduced manifold vacuum allows the spring on diaphragm D to move the diaphragm downward. Lever E will move upward under valve F and force the valve open. When valve F is opened,

Types of Propane Conversion Equipment

FIGURE 3-3. EC1-Equipped Vaporizer. *(Courtesy of Impco Carburetion, Inc.)*

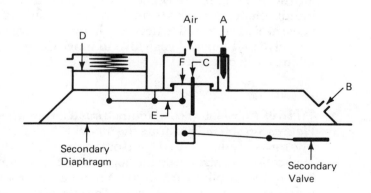

FIGURE 3-4. EC1 Internal Design.

atmospheric pressure is applied to the secondary diaphragm and air-fuel ratios become richer. Idling the engine causes the secondary diaphragm to hit screw C and open valve F. Bleed screw A becomes ineffective when valve F is open. (Adjustment of EC1 equipment is discussed in Chapter 5.)

MIXER DESIGN Impco propane mixers operate on the principle of the tapered gas valve (item V in Figure 3-5), which is attached to the bottom of the air valve. When the engine is not operating, spring S seats the tapered gas valve in the vapor inlet. Vacuum passage P allows the vacuum from the air valve area to be sensed on top of the air valve. Diaphragm D allows vertical movement of the air valve. Component I is an idle mixture screw. Air flow past the mixture screw bypasses the air valve area. A full load screw, component A, is used to adjust air-fuel mixtures at wide open throttle.

MIXER OPERATION When the engine is cranking, the rush of air through the mixer creates a slight vacuum in the air valve area of the mixer. The vacuum is applied through passage P to the upper side of the air valve and diaphragm. The air valve and tapered gas valve will be lifted when the vacuum is applied to the diaphragm. Propane vapor will flow past the tapered gas valve into the air stream and cause the engine to start. The vacuum in the air valve area increases in relation to engine rpm. The increase in vacuum lifts the tapered gas valve as engine speed increases. Correct air-fuel ratios are maintained by the design of the tapered gas valve. Throttle opening controls engine rpm by regulating the amount of air and propane vapor entering the cylinders. During wide open throttle operation, the tapered gas valve is lifted out of the vapor inlet, and the air-fuel mixture is determined by the full load screw adjustment. When the idle mixture screw is turned outward, additional air will bypass the air valve area. The resulting decrease in the air valve vacuum will cause the air valve and tapered gas valve to move downward and thus a leaner mixture will be created.

TYPES OF MIXERS Impco manufactures a large variety of propane mixers. Figures 3-6 and 3-7 illustrate the model 125 and 225 mixers, respectively. The mixer is adapted to the top of the gasoline carburetor for a dual fuel application. Straight propane conversions have the mixer adapted to the base of the gasoline carburetor. A new carburetor base may be supplied for straight propane conversions. The idle mixture screw is item A and the full load screw is component B in Figures 3-6 and 3-7. Arrows or index

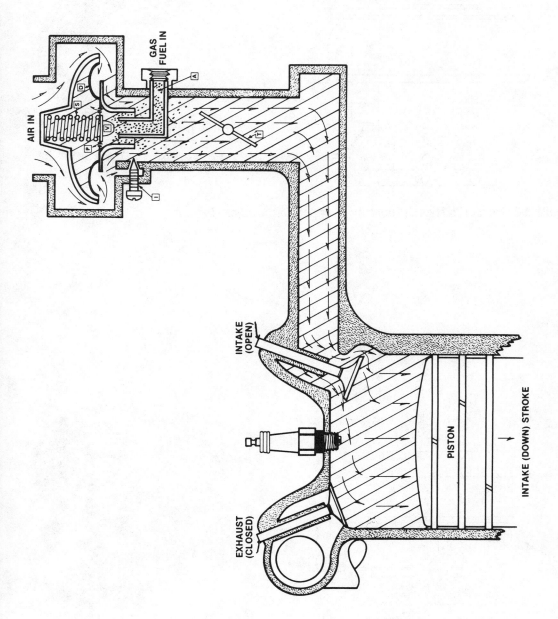

FIGURE 3-5. Propane Mixer. *(Courtesy of Impco Carburetion, Inc.)*

Alternate Automotive Fuels

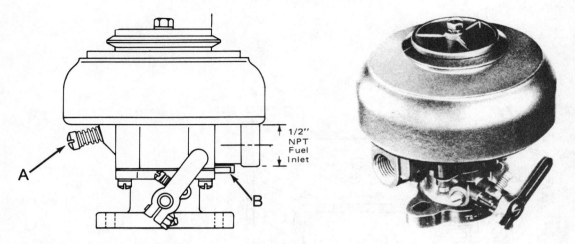

FIGURE 3-6. Model 125 Mixer. *(Courtesy of Impco Carburetion, Inc.)*

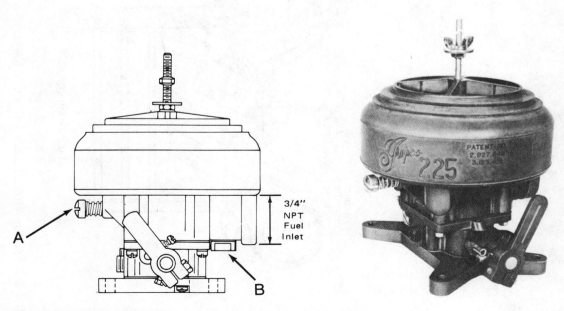

FIGURE 3-7. Model 225 Mixer. *(Courtesy of Impco Carburetion, Inc.)*

Types of Propane Conversion Equipment

marks around the mixture screws indicate the direction of rotation to obtain rich or lean air-fuel mixtures.

An Impco model 300 mixer is pictured in Figure 3-8. Two sizes of 300 mixers are available, a 300-20 and a 300-50. A dual fuel application is shown in Figure 3-8 with the mixer connected to the gasoline carburetor adapter.

The idle mixture screw and the power mixture screw on the 300 mixer are illustrated in Figure 3-9. The idle mixture screw is located under a plug on top of the vapor inlet. Clockwise rotation of the idle mixture screw turns the tapered gas valve downward and makes the idle air-fuel ratio leaner. The idle mixture screw is attached to the tapered gas valve. Index marks on the power screw indicate which direction the screw should be rotated to obtain lean or rich mixtures.

Figure 3-10 shows an Impco 425 mixer for a straight propane application. The large screw located on the front of the mixer adjusts full load or power mixtures. Idle mixture is corrected by turning the small screw on the side of the mixer. The effect of this adjustment is indicated by an arrow around the mixture screw.

In dual fuel installations the 425 mixer is usually mounted horizontally, and an adapter connects the mixer to the gasoline carburetor, as indicated in Figure 3-11.

FIGURE 3-8. Model 300 Mixer. *(Courtesy of Impco Carburetion, Inc.)*

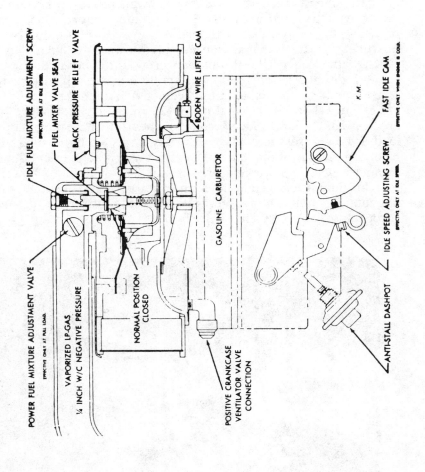

FIGURE 3-9. Model 300 Mixer Adjusting Screws. *(Courtesy of Impco Carburetion, Inc.)*

Types of Propane Conversion Equipment

FIGURE 3-10. Model 425 Mixer Straight Propane Application. *(Courtesy of Impco Carburetion, Inc.)*

MIXER SELECTION Propane mixers must be selected according to the cubic feet per minute (cfm) of air required by the engine. The mixer cfm rating must be matched to the engine cfm requirements. The cfm chart in Table 3-1 indicates engine cfm requirements at various speeds. A 350 cubic inch displacement (cid) engine operating at 4,000 rpm, for example, requires 343 cfm.

Mixer cfm ratings are presented in Table 3-2. An engine with a 343 cfm requirement would need a 300-50 mixer rated at 432 cfm. Insufficient mixer cfm will result in reduced air flow and loss of power at high speed.

FIGURE 3-11. Model 425 Mixer Dual Fuel Application. *(Courtesy of Impco Carburetion, Inc.)*

VACUUM FUELOCK A vacuum fuelock may be used to turn the liquid propane flow off and on between the fuel tank and the vaporizer. The fuel inlet is located in the top cover of the Impco model VFF 30 fuelock, as indicated in Figure 3-12. Fuel exits through the larger opening in the side of the fuelock. The mixer vacuum is connected to the small outlet in the side of the VFF 30. The manifold vacuum may be connected to the fuelock if cold starting is difficult.

Internal construction of the vacuum fuelock is shown in Figure 3-13. The vacuum is applied between the diaphragm and the center section of the VFF 30. Cranking the engine creates a vacuum in the fuelock that lifts the diaphragm and lever upward. Vertical movement of the lever opens the outlet valve and lets propane flow through the fuelock. When the engine is stopped, the fuelock valve remains closed and propane is unable to flow.

TABLE 3-1. Engine cfm Requirements

(CUBIC FEET/MINUTE)

Naturally aspirated air flow figures at 85% volumetric efficiency. For 4-cycle gasoline and diesel. Double the CFM figure for 2-cycle engines.

ENGINE RPM (REVOLUTIONS PER MINUTE)

C.I.D.	400	600	800	1000	1200	1400	1600	1800	2000	2200	2400	2600	2800	3000	3200	3400	3600	3800	4000
100	11	15	19	24	30	35	39	44	49	54	59	63	69	74	79	83	89	93	98
150	15	22	30	37	44	52	59	66	74	81	89	96	103	110	118	124	132	139	147
200	20	30	39	49	59	69	78	89	98	108	116	128	137	147	156	167	176	187	195
250	24	37	50	61	74	82	98	111	122	135	147	159	172	184	196	209	220	232	245
300	30	44	59	74	89	103	118	133	148	162	177	192	207	219	236	249	264	279	293
350	34	52	69	85	103	120	137	154	171	188	206	224	241	257	274	291	309	326	343
400	39	59	78	98	117	137	156	176	195	215	235	255	275	295	314	334	353	373	392
450	44	66	88	115	136	157	178	199	220	243	265	287	309	331	353	377	399	421	442
500	49	89	98	122	147	171	195	219	245	269	285	319	344	368	394	418	443	467	492
550	54	80	108	135	161	187	214	243	270	296	323	350	379	407	436	464	487	515	540

SOURCE: Courtesy Impco Carburetion Inc.

TABLE 3-2. Mixer cfm Ratings

Model		Cubic Feet/Minute
50–500	@ 67 HP	108
125	@ 126 HP	202
225	@ 205 HP	329
300A-1, 300A-20	@ 217 HP	348
300A-50, 300A-70	@ 270 HP	432
425	@ 287 HP	460

SOURCE: Courtesy Impco Carburetion Inc.

FIGURE 3-12. Model VFF 30 Fuelock. *(Courtesy of Impco Carburetion, Inc.)*

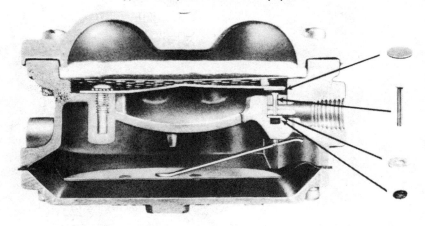

FIGURE 3-13. Model VFF 30 Fuelock. *(Courtesy of Impco Carburetion, Inc.)*

Garretson Propane Equipment

VAPORIZER DESIGN The internal structure of Garretson vaporizers is very similar to that of the Impco equipment described in the preceding pages. An electric fuelock is mounted to the liquid propane inlet. The primary seat is accessible by removing the fuel inlet nut. Adjustment of the secondary valve spring is accomplished by turning the slotted screw below the fuel inlet fitting. The adjusting screw is secured by a lock plug. Clockwise rotation of the adjusting screw tightens the secondary valve spring and makes the air-fuel mixture leaner. The vapor outlet is located at the bottom of the vaporizer. An idle attachment is threaded into the vaporizer above the vapor outlet. Idle fuel is taken from the primary section at 3 psi and is fed through the idle attachment to the mixer. The idle control unit may be mounted on the mixer. Clockwise rotation of the idle mixture screw makes the air-fuel mixture leaner. Coolant inlet and outlet fittings are provided in the vaporizer. As Figure 3-14 illustrates, a fuel filter is mounted beside the vaporizer and is connected to the fuelock.

The electric prime solenoid in the back of the vaporizer (Figure 3-15) pushes the secondary diaphragm and opens the secondary valve. The prime solenoid is activated automatically by the starting motor. It is also energized when the dash control switch is placed in the manual prime position.

FIGURE 3-14. Vaporizer Design. *(Courtesy of Garretson Equipment Company, Inc.)*

FIGURE 3-15. Vaporizer Prime Solenoid. *(Courtesy of Garretson Equipment Company, Inc.)*

The amount of vapor flow while priming can be adjusted by rotating the plastic nut on the end of the solenoid plunger. The plunger must be held stationary when the nut is being adjusted. Clockwise rotation of the adjusting nut increases the amount of prime vapor flow. The prime solenoid is adjusted correctly when the engine almost stalls at idle if the plunger is pushed in manually. A plastic cap protects the prime solenoid plunger. Complete wiring diagrams are provided in Chapter 4.

MIXER DESIGN Garretson mixers utilize a venturi ring connected to the vapor inlet. Vapor discharge holes are spaced around the venturi ring, and a venturi plate centered in a flexible diaphragm rests on top of the venturi ring. The venturi plate diaphragm is sandwiched between the air cleaner cover and the air filter. Adjustable venturi plate screws determine the venturi opening between the plate and the ring, as illustrated in Figure 3-16.

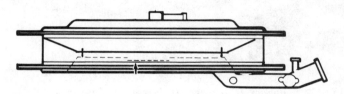

FIGURE 3-16. Mixer Design. *(Courtesy of Garretson Equipment Company, Inc.)*

The full load mixture screw is located in the side of the vapor inlet. Garretson mixers are available in three sizes to fit various engines. A model 212-01 mixer fits engines up to 300 CID. The model 213-01 mixer is available for engines up to 350 CID that have two-barrel carburetors. Engines from 300 CID to 500 CID with four-barrel carburetors should be fitted with a model 214-01 mixer. Adapters are available to connect the mixer to the top of the gasoline carburetor. The mixer may be adapted to the base of the carburetor in a straight propane application. In straight propane conversions, the gasoline fuel pump is removed. A vacuum hose is connected from the vapor inlet to the top of the air cleaner cover, as illustrated in Figure 3-17. In the gasoline mode, the venturi plate and diaphragm will be lifted as the vacuum in the vapor inlet increases slightly. When propane fuel is used, the venturi plate rests on the venturi ring.

FIGURE 3-17. Mixer Design. *(Courtesy of Garretson Equipment Company, Inc.)*

MIXER ADJUSTMENT The vacuum between the venturi plate and ring is determined by the venturi opening. Adjustment of the venturi plate screws provides correct air-fuel ratios. Initial settings for venturi plate screws are listed in Table 3-3. (Detailed adjustment of air-fuel ratios is provided in Chapter 5.)

IDLE CONTROL VALVE The idle control valve may be mounted on the mixer or the vaporizer. Idle fuel flows from the vaporizer primary section through the idle control valve to the mixer venturi ring. Idle air-fuel ratios are adjusted by rotating the mixture screw in the idle control valve. Leaner mixtures are obtained by rotating the idle mixture screw clockwise. A hose from the intake manifold is connected to the idle control valve. When the engine is cranking or running, the manifold vacuum opens the idle control valve. The idle control valve closes when the engine is shut off and thus prevents any further flow of propane vapor. On a Garretson system mixer, the vacuum will not open the secondary vaporizer valve during idle operation. The idle control valve supplies idle fuel requirements. Accelerating the engine off idle increases

TABLE 3-3. Venturi Plate Adjustment

Cu. In. Disp.	Height Required	Screw Turns From Flush
Set up for models no. 212 & no. 213 vac-u-lift mixers		
75	.0283	1/8
85	.0330	1/4
100	.0381	1/2
110	.0435	5/8
120	.0462	3/4
150	.0613	1
175	.0720	1-1/4
200	.0795	1-1/2
225	.0916	2
250	.1001	2-1/4
275	.1091	2-1/2
300	.1231	3
325	.1380	3-1/4
350	.1431	3-1/2
375	.1484	3-3/4
400	.1591	4
Set up for model 214 vac-u-lift II gas/air mixer		
250		1
300		1-1/2
350		1-3/4
400		2
450		2-1/2
500		3

SOURCE: Courtesy Garretson Equipment Co. Inc.

FIGURE 3-18. Idle Control Valve. *(Courtesy of Garretson Equipment Company, Inc.)*

the mixer vacuum, and the secondary vaporizer valve opens to supply additional fuel. A plastic pin on the idle control valve diaphragm extends through the valve cover, as illustrated in Figure 3-18. The plastic pin is lifted to determine whether the idle control valve is open. The valve should be open while the engine is being cranked.

Tartarini Propane Equipment

VAPORIZER DESIGN Tartarini vaporizers consist of a three-stage unit. The vapor outlet is the large connection at the top of the vaporizer in Figure 3-19. Heater hoses are connected to the fittings positioned on the right

FIGURE 3-19. Tartarini Vaporizer. *(Courtesy of Ontario Ministry of Transportation and Communication)*

side of the vaporizer. Liquid propane enters the inlet fitting at the bottom of the unit. The prime solenoid is located on the left side of the vaporizer.

Idle adjustment is accomplished by turning the large screw at the top of the vaporizer. Two unique features of this unit are the power valve and the shutdown diaphragm. The power valve is component 19 in Figure 3-20, and the shutdown diaphragm is item 14.

VAPORIZER OPERATION When the fuelock is energized, liquid propane flows through the gas inlet and past the first-stage seat (item 4 in Figure 3-20) into area A. As the propane flows into area A, it changes to a vapor because of the reduction in pressure. Vapor pressure of 22 psi (154 kp) is required to push the first-stage diaphragm upward and close the seat, item 4. Propane vapor flows from area A past the second-stage seat, item 6, into area B. When vapor pressure in area B reaches 5 psi (35 kp), the second-stage diaphragm is pushed upward and the seat, item 6, is closed. Vapor enters area C through the third-stage seat, item 8. When the engine is running, the mixer vacuum applied to area C moves the third-stage

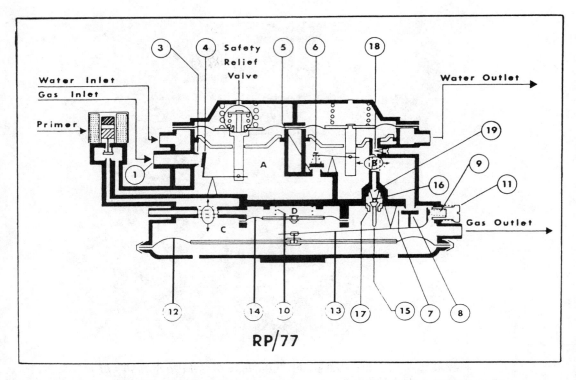

FIGURE 3-20. Tartarini Vaporizer Internal Design. *(Courtesy of Alternative Fuel Systems, Ltd.)*

diaphragm, item 12, upward. Vertical movement of the third-stage diaphragm increases the third-stage valve opening and additional vapor is able to flow to the mixer. Third-stage diaphragm movement and the mixer vacuum increase directly in relation to engine rpm. Vent openings in the third-stage diaphragm cover allow atmospheric pressure to be exerted between the diaphragm and cover. When propane vapor is moved out of area A or B, the diaphragm springs are allowed to push the diaphragms down and open their respective seats. When the valves open, vapor refills area A or B to the pressures mentioned above.

The idle mixture screw is item 11. Clockwise rotation of the screw pushes leaf spring 13 upward, closing the third-stage valve and creating a leaner air-fuel ratio.

A clearance of 0.025 (0.635 mm) exists between spring 13 and the power valve stem, item 15, when the engine is not running. As the throttle approaches the wide open position, the third-stage diaphragm and leaf spring will move upward, pushing the power valve stem open. The mixture becomes enriched as propane vapor flows from the second-stage area past the power valve into the gas outlet. Power valve enrichment is preset by the adjusting screw, item 18. Power valve action is illustrated in Figure 3-21.

The manifold vacuum is connected to area D. While the engine is running, diaphragm 14 is held away from the third-stage diaphragm. When the engine is not operating, spring 10 will push diaphragm 14 against the third-stage diaphragm and close the third-stage valve. The action of diaphragm 14 prevents any flow of propane vapor into the third stage if the engine is stopped.

When the prime solenoid is energized, vapor can flow from the first stage directly into the third stage to the gas outlet. If the first-stage seat becomes defective, the safety relief valve limits pressure in area A to 88 psi (616 kp). Area B could function safely at 88 psi (616 kp), and the third-stage operation would be satisfactory.

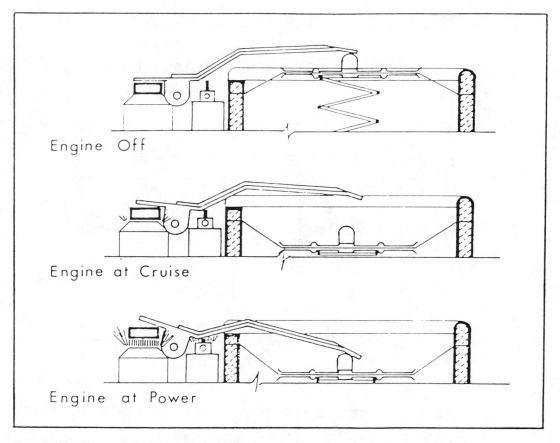

FIGURE 3-21. Tartarini Power Valve Action. *(Courtesy of Alternative Fuel Systems, Ltd.)*

MIXER DESIGN Propane vapor is discharged from a series of holes positioned around the venturi ring, as illustrated in Figure 3-22. A venturi plate and flexible diaphragm are positioned above the venturi ring. Fixed spacer rivets determine the distance between the venturi ring and plate. The venturi opening is not adjustable. Early Tartarini systems required a full load screw in the vapor hose between the vaporizer and the mixer, but in newer mixers the size of the venturi opening and discharge holes is computer-matched to the engine CID, and the full load screw is not required.

MIXER OPERATION When propane fuel is used, the venturi plate is always positioned above the venturi ring, as indicated in the left half of Figure 3-22. If the gasoline mode is selected, the manifold vacuum is applied to

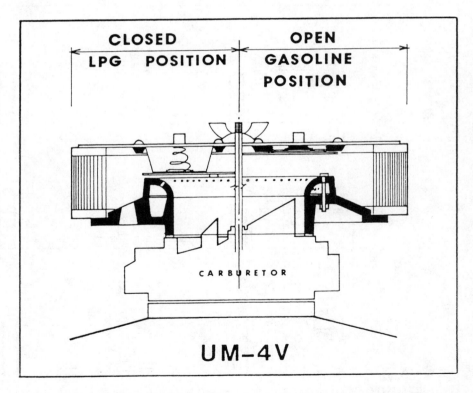

FIGURE 3-22. Tartarini Dual Fuel Mixer. *(Courtesy of Alternative Fuel Systems, Ltd.)*

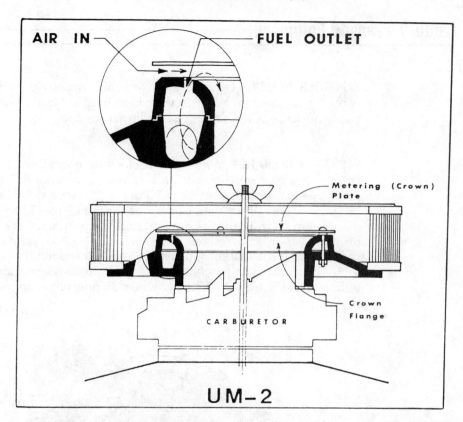

FIGURE 3-23. Tartarini Straight Propane Mixer. *(Courtesy of Alternative Fuel Systems, Ltd.)*

the flexible diaphragm. The venturi plate is lifted up to provide unrestricted air flow to the gasoline carburetor, as shown in Figure 3-22. An electric solenoid energized by a dash switch applies the vacuum to the flexible diaphragm. Under low vacuum conditions in the gasoline mode, a check valve in the vacuum hose prevents the venturi plate from dropping down. On straight propane applications, a fixed venturi plate replaces the movable plate, as illustrated in Figure 3-23. The mixer is designed to fit on top of the gasoline carburetor. The gasoline fuel pump must be removed on straight propane applications.

Century Propane Equipment

VAPORIZER DESIGN Century vaporizers are two-stage units. Fuel and coolant fittings in the vaporizer are identified in Figure 3-24. An optional vapor outlet is available for use on larger engines.

VAPORIZER OPERATION When the fuelock is energized, liquid propane enters the vaporizer fuel inlet and changes to a vapor as it flows past the primary seat, component B in Figure 3-25. Area C is then filled with propane vapor. When vapor pressure reaches 5 psi (35 kp), the primary diaphragm, item D, moves upward and the primary valve closes. When propane vapor is moved out of the primary section, primary pressure is reduced, and the diaphragm spring is able to reopen the primary valve.

Mixer vacuum is applied to area V in the secondary section, as indicated in Figure 3-26. The secondary diaphragm, component H, will be

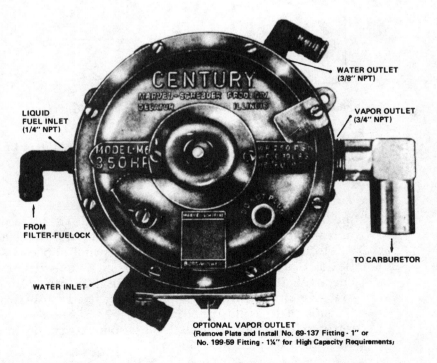

FIGURE 3-24. Century Vaporizer. *(Courtesy of Borg Warner Corporation)*

Types of Propane Conversion Equipment

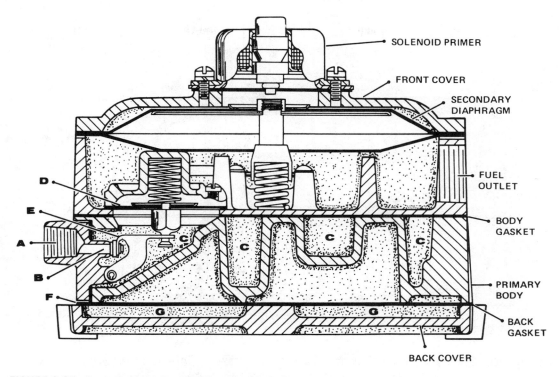

FIGURE 3-25. Century Vaporizer Primary Section. *(Courtesy of Borg Warner Corporation)*

moved downward by the mixer vacuum. The secondary valve, component J, will be opened by downward movement of the secondary diaphragm and lever, and propane vapor will be drawn from the primary section through the secondary valve to the vapor outlet and mixer. Mixer vacuum and secondary diaphragm movement are proportional to engine speed. Atmospheric pressure is available between the secondary diaphragm and cover. When the engine is shut off, the secondary valve will close, thereby preventing any further movement of propane vapor. Coolant is circulated through area W to assist in vaporizing the liquid propane. The prime solenoid in the secondary diaphragm cover forces the secondary valve open, and provides fuel vapor to the mixer for easier cold starting.

A transfer tube installed between the primary diaphragm cover and the secondary valve is illustrated in Figure 3-27. As engine demand increases, fuel flow through the secondary valve creates a pressure drop in the transfer tube and the atmospheric side of the primary diaphragm. Reduced pressure on top of the primary diaphragm allows the

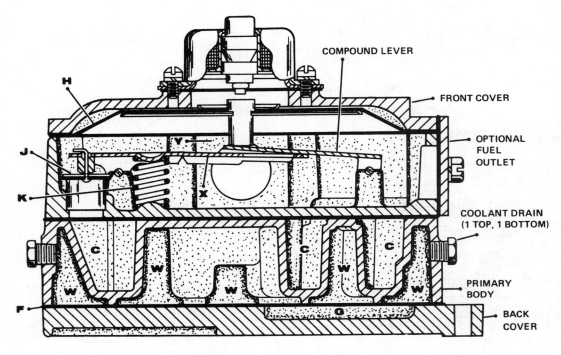

FIGURE 3-26. Century Vaporizer Secondary Section. *(Courtesy of Borg Warner Corporation)*

diaphragm to move faster. A more constant primary pressure is achieved under all engine load conditions.

MODEL 2610 MIXER The Century 2610 mixer is designed to be adapted to the top of various gasoline carburetors. Propane vapor enters the vapor inlet, item A in Figure 3-28. Passages in the mixer body conduct the fuel vapor to the bottom of the two fuel doors located on each side of the mixer. One of the fuel doors can be seen in item B in Figure 3-28. Propane vapor flows through the hollow fuel doors and discharges from a slot in the top of the doors. Linkage C connects the fuel door to a tapered gas cone in the mixer. Fuel vapor flow through the mixer is metered by the tapered gas valve. Because the fuel doors are linked to the tapered gas valve, the flow of fuel can be regulated precisely in relation to the flow of air.

Simultaneous opening of the fuel doors is ensured by a linkage between the doors. The air velocity through the mixer opens the fuel

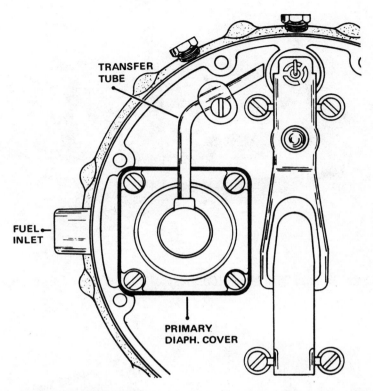

FIGURE 3-27. Century Vaporizer Transfer Tube. *(Courtesy of Borg Warner Corporation)*

doors against the tension of the closing spring in relation to engine speed. Free movement of the fuel doors is essential. The doors must snap closed from the open position. Fuel door action should be checked by pushing simultaneously on both doors. A power valve, item D, provides mixture enrichment under heavy load conditions. The power valve is controlled by the manifold vacuum. The idle air-fuel ratio is adjusted by rotating the shaft between the tapered gas valve and the fuel door. The gas valve shaft can be rotated by inserting a steel rod in the hole provided in the shaft. Fuel door linkages are illustrated in Figure 3-29. A bowden cable from the fuel door linkage to a dash control lever allows the fuel doors to be opened manually when the engine is operated in the gasoline mode. A full load screw is installed in the vaporizer vapor outlet connected to the 2610 mixer.

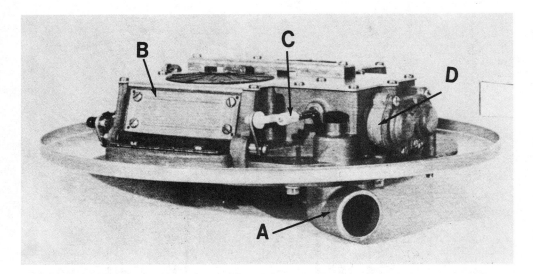

FIGURE 3-28. Century 2610 Mixer. *(Courtesy of Borg Warner Corporation)*

FIGURE 3-29. Century 2610 Mixer. *(Courtesy of Borg Warner Corporation)*

Vialle Propane Equipment

VAPORIZER DESIGN The Vialle vaporizer is a two-stage unit that varies somewhat in its internal operation. Figure 3-30 shows a Vialle vaporizer. The vapor outlet is the large fitting near the top of the unit. The fuel inlet is located near the bottom right side of the vaporizer. Coolant connections and the idle mixture screw are pictured on the right side of the unit. The solenoid near the top of the diagram is referred to as a fuel shutoff solenoid.

FIGURE 3-30. Vialle Vaporizer. *(Courtesy of Vialle Autogas Systems)*

VAPORIZER OPERATION When the fuelock is energized, liquid propane flows past the primary valve and seat, and as it enters the primary section it is changed to a vapor owing to the pressure reduction. The primary valve is shown in the open position in Figure 3-31.

Under primary pressure of 5 psi, the primary diaphragm will move upward, and the primary valve will close, thus preventing any further fuel flow, as indicated in Figure 3-32. When propane is used out of the primary section, the primary spring pushes the diaphragm downward. The primary valve is then opened by downward movement of the diaphragm, and fuel is allowed to flow into the primary section. Coolant circulation through the vaporizer assists in fuel vaporization.

Fuel enters the secondary section from the primary area. When the engine is idling, the secondary valve is closed and propane vapor is routed past the idle mixture screw and through the vapor outlet. Some fuel vapor will also flow through orifice 3 into tube 6 to the vapor outlet, as illustrated in Figure 3-33. A slight vacuum is developed between orifice 3 and tube 6. At idle speed, the vacuum at orifice 3 is insufficient to move diaphragm 4 against its seat. With diaphragm 4 unseated, the vacuum in area 5 is insignificant and the secondary valve remains closed.

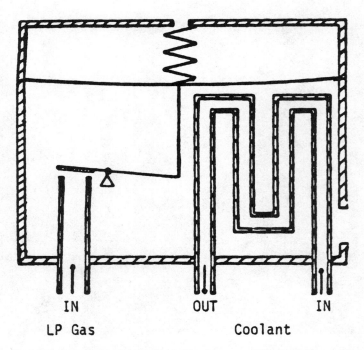

FIGURE 3-31. Vaporizer Primary Section. *(Courtesy of Vialle Autogas Systems)*

Types of Propane Conversion Equipment

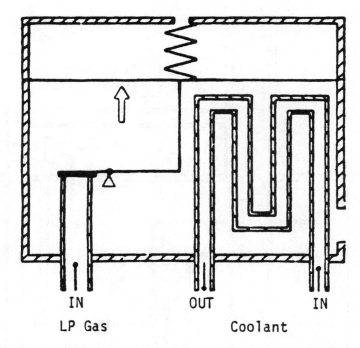

FIGURE 3-32. Vaporizer Primary Section. *(Courtesy of Vialle Autogas Systems)*

The propane fuelock and the vaporizer shutoff valve are both energized by an electronic relay. When the shutoff valve (item 1 in Figure 3-33) is energized, idle fuel is allowed to flow through the vaporizer. When the ignition switch is turned on in the LPG mode, battery voltage will be supplied to the propane fuelock and the vaporizer shutoff valve from the electronic relay, as illustrated in Figure 3-34.

Propane vapor will flow through the vaporizer idle passages to the mixer for priming purposes. If no attempt is made to start the engine, the electronic relay opens the circuit to the fuelock and the shutoff solenoid in 2–3 seconds. A Vialle system can be primed by turning on the ignition switch and waiting 2 or 3 seconds before starting the engine. A wiring connection from the ignition coil signals the electronic relay when the engine is running. The electronic relay supplies battery voltage to the shutoff solenoid and the fuelock when the engine is cranking or running. If the engine stalls with the ignition switch on, the electronic relay opens the circuit to the fuelock and shutoff valve and thus prevents any further fuel flow.

When the throttle is opened from the idle position, the increase in the mixer vacuum allows the control diaphragm 4 to be seated on the

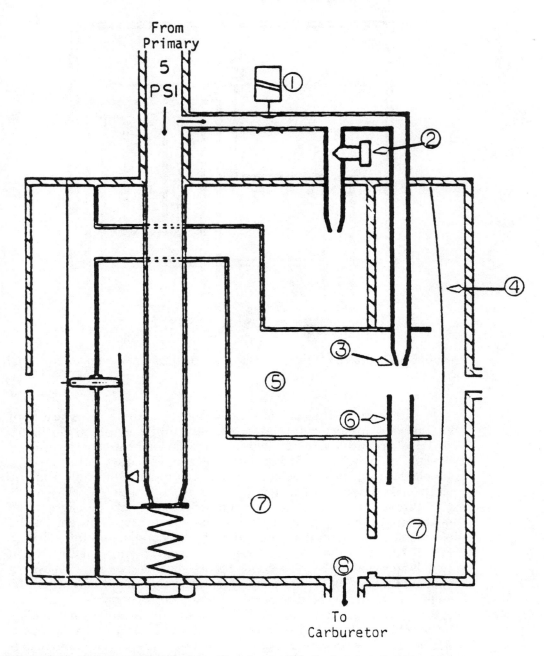

FIGURE 3-33. Vaporizer Idle Operation. *(Courtesy of Vialle Autogas Systems)*

Types of Propane Conversion Equipment   55

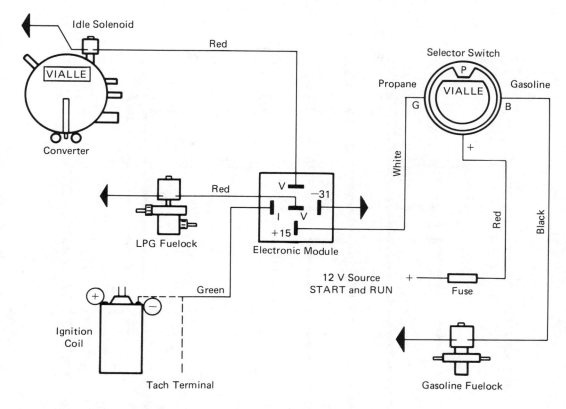

FIGURE 3-34. Electrical Circuit. *(Courtesy of Vialle Autogas Systems)*

opening of passage 5, as illustrated in Figure 3-35. As a result, the vacuum between orifice 3 and port 6 is applied to the secondary diaphragm, which is moved by the increased vacuum in passage 5 and by the atmospheric pressure between the secondary diaphragm and cover. Movement of the secondary diaphragm opens the secondary valve and allows additional fuel vapor to flow past the secondary valve to the mixer. The vacuum between orifice 3 and port 6 increases in relation to engine rpm. The secondary valve opening and fuel flow will be directly proportional to the vacuum in passage 5 and engine rpm.

VAPOR PURGE SYSTEM Vaporizer diaphragm movement and air-fuel mixtures may be affected by oil in the propane collecting in the vaporizer. The Vialle vapor purge system (VPS) eliminates this problem. The VPS utilizes a vacuum reservoir and a sludge collector. One vacuum reservoir

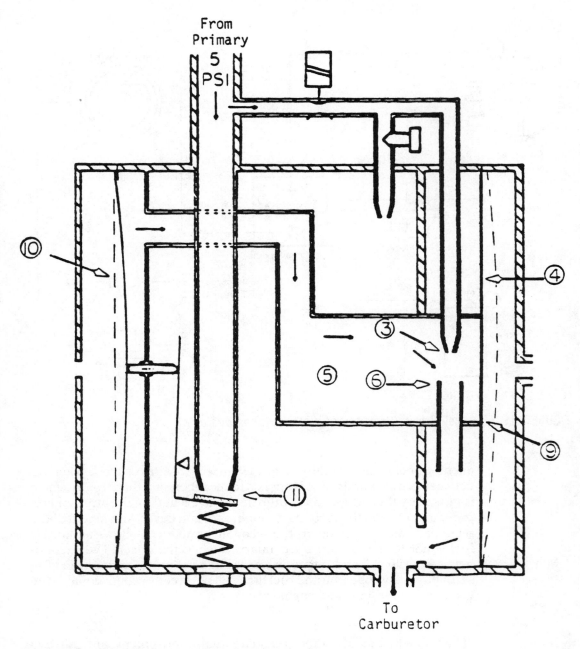

FIGURE 3-35. Vaporizer Secondary Section. *(Courtesy of Vialle Autogas Systems)*

outlet is connected to the intake manifold. A vacuum hose connects the vacuum reservoir to the sludge collector. The second outlet on the sludge collector is connected to the vaporizer drain connection, as indicated in Figure 3-36.

When the engine is running, the vacuum is contained in the vacuum reservoir by an electrically operated nonreturn valve. Switching the ignition off deenergizes the nonreturn valve and allows the manifold vacuum to enter the sludge collector. Oil sludge from the vaporizer is moved into the sludge collector by the manifold vacuum.

MIXERS Vialle propane mixers are available to fit a wide variety of applications. A series of vapor discharge holes are spaced around the venturi ring in the mixer. A full load screw is located in the vapor hose between the vaporizer and the mixer. The venturi ring may be located inside the gasoline air cleaner, as shown in Figure 3-37. On some applications the mixer ring may be installed between the gasoline air cleaner and the carburetor, as pictured in Figure 3-38.

The propane mixer may be positioned between the carburetor bowl and throttle body, as shown in Figure 3-39. The mixer may also be located in the air cleaner intake hose, as illustrated in Figure 3-40.

Deactivation of the vacuum-operated air cleaner heat control valve is essential when an engine is operating on propane fuel. An electrically operated vacuum solenoid should be installed in the vacuum hose to the air cleaner, as pictured in Figure 3-41. The vacuum solenoid can be wired

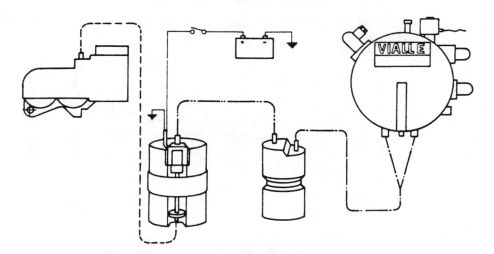

FIGURE 3-36. Vapor Purge System. *(Courtesy of Vialle Autogas Systems)*

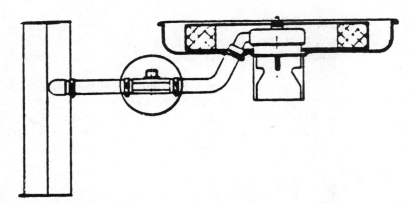

FIGURE 3-37. Mixer Installation in the Air Cleaner. *(Courtesy of Vialle Autogas Systems)*

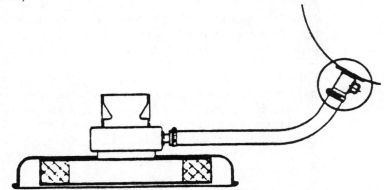

FIGURE 3-38. Mixer Installation Under the Air Cleaner. *(Courtesy of Vialle Autogas Systems)*

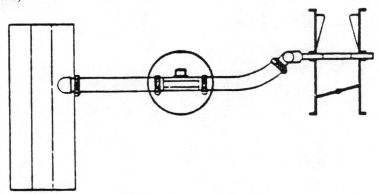

FIGURE 3-39. Mixer Installation Between the Carburetor Float Bowl and Base. *(Courtesy of Vialle Autogas Systems)*

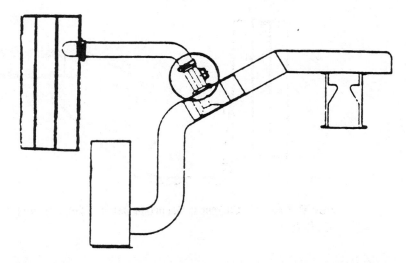

FIGURE 3-40. Mixer Installation in the Intake Hose. *(Courtesy of Vialle Autogas Systems)*

to the gasoline fuelock. In the propane mode, the gasoline fuelock and the vacuum solenoid are deenergized. The vacuum to the air cleaner diaphragm is shut off by the solenoid, and thus the air cleaner butterfly door is allowed to remain open. Cold air is drawn into the air cleaner when the butterfly door is open.

On some applications a balance tube may have to be installed between the vaporizer secondary diaphragm vent and the air cleaner, as outlined in Figure 3-42. The balance tube creates a balance in pressure between the air cleaner and the vented side of the secondary diaphragm. The two pressures are balanced to compensate for any air cleaner restriction and to provide more stable air-fuel mixtures.

FIGURE 3-41. Air Cleaner Vacuum Door Operation. *(Courtesy of Vialle Autogas Systems)*

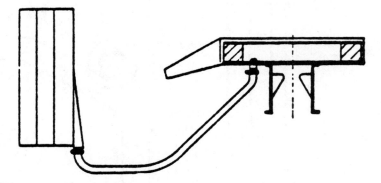

FIGURE 3-42. Air Cleaner to Vaporizer Balance Tube. *(Courtesy of Vialle Autogas Systems)*

Petrosystems Propane Equipment

MIXERS The Petrosystems mixer is installed on top of the gasoline carburetor in a dual fuel system. Adapters are available to connect the mixer to the top of most gasoline carburetors. In a straight propane conversion, the mixer is attached to the base of the gasoline carburetor by means of a special adapter. The float bowl and air horn assemblies in the gasoline carburetor are discarded in a straight propane conversion. There is only one moving part in the air-flow-controlled constant velocity mixer illustrated in Figure 3-43.

SWITCHING VALVE The switching valve is mounted near the air-fuel mixer. A dash-mounted control lever and bowden cable operate the switching valve. The gasoline line to the carburetor is connected to the switching valve. Gasoline flow to the carburetor and the vacuum to the vaporizer are switched on and off by the switching valve shown in Figure 3-44.

The manifold vacuum is connected to one port on the switching valve, and the other port is connected to the vaporizer, as pictured in Figure 3-45. In the gasoline mode, the switching valve turns on the gasoline flow to the carburetor and switches off the vacuum to the vaporizer. When the dash control lever is positioned in the LPG mode, the gasoline flow is shut off, and the vacuum is turned on to the vaporizer. A

Types of Propane Conversion Equipment 61

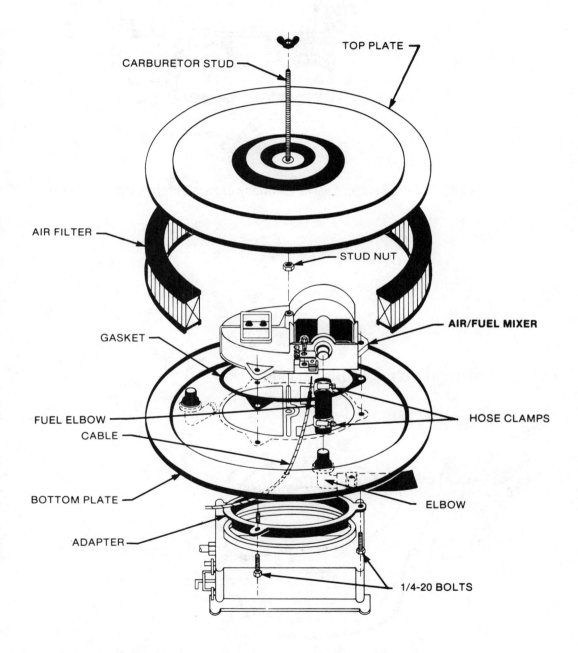

FIGURE 3-43. Air-Fuel Mixer. *(Courtesy of Petrosystems International, Inc.)*

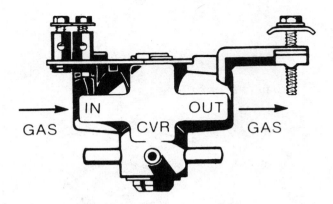

FIGURE 3-44. Switching Valve. *(Courtesy of Petrosystems International, Inc.)*

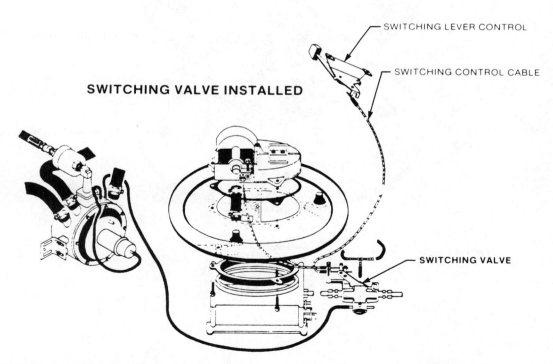

FIGURE 3-45. Switching Valve Installation. *(Courtesy of Petrosystems International, Inc.)*

vacuum fuelock is located in the vaporizer. Even a slight cranking vacuum will open the propane fuelock and allow liquid propane to flow to the primary section of the vaporizer.

As well as operating the switching valve, the dash control lever pulls the mixer open in the gasoline mode. Cable adjustments are important to proper system operation. Sharp bends or kinks in the cables must be avoided. When the cables have been installed, the adjustment should proceed as follows:

1. Adjust switching valve to dash lever cable so the switching valve lever contacts its stop when the dash lever is in the LPG position.
2. Adjust switching valve to mixer cable so the slider in the mixer is just closed with the switching valve lever against its stop and the idle mixture screw backed out so it allows the slider to close.
3. Check free movement of dash-control lever from LPG to gasoline modes.

CONVERTERS The converter, or vaporizer, is similar to the units used by other manufacturers. An external in-line filter is mounted on top of the unit. Heater hoses are connected to the vaporizer to assist in converting liquid propane to a vapor. (Correct heater hose connections are discussed in the next chapter.) As mentioned earlier, an integral vacuum fuelock is used in the vaporizer. The manifold vacuum applied to the diaphragm cover port controls the flow volume of propane vapor. The Petrosystems converter is shown in Figure 3-46.

Propane Diesel Boosting Equipment

McCOY MILEAGE MASTER Propane vapor may be injected into the air intake on diesel engines to increase horsepower. The propane diesel boost system shown in Figure 3-47 withdraws vapor from the propane fuel tank.

Propane vapor flows through the filter and solenoid to the pressure regulator. Tank pressure is reduced to less than 1 psi (7 kp) by the regulator. Fuel flows from the regulator through the controlled metering valve to the fuel injection tube in the air intake. The metering valve rod must be clamped to the diesel injection pump-operating rod so the metering valve is closed at idle speed. The opening of the metering valve increases in relation to engine rpm. The fuel injection tube should be

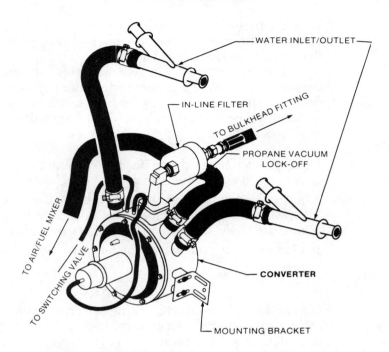

FIGURE 3-46. Petrosystems Converter External Connections. *(Courtesy of Petrosystems International, Inc.)*

positioned in the air intake as illustrated in Figure 3-47. A dash control switch is used to energize the fuelock solenoid. Some manufacturers of propane diesel-boosting equipment recommend recalibration of the diesel injection pump to a lower setting when a propane booster is installed.

──────────────── **Questions** ────────────────

1. Liquid propane changes to a vapor in the secondary section of the vaporizer.   T F
2. In a two-stage vaporizer the secondary section is operated by _____ _____.
3. The Impco EC1-equipped vaporizer allows _____ air-fuel ratios.
4. Mixer cfm rating must be matched to the _____.

Types of Propane Conversion Equipment

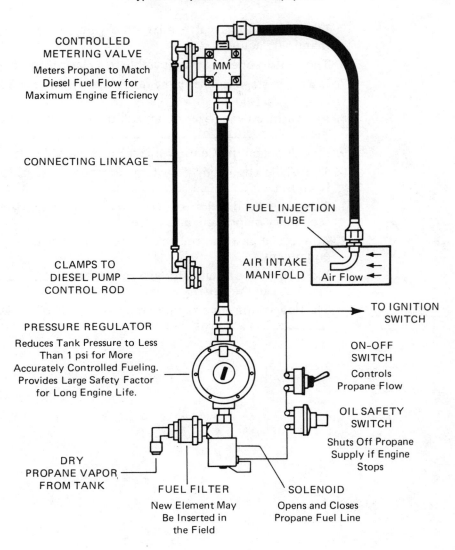

FIGURE 3-47. Propane Diesel Boost System.

5. In a tapered gas valve mixer the valve is opened by the intake manifold vacuum. T F
6. The venturi opening is adjustable in a Garretson mixer. T F
7. In a three-stage vaporizer the first-stage pressure would be _____ _____ psi.
8. In a Tartarini vaporizer richer full-throttle mixtures are provided by the _____ _____.
9. Fuel doors on the Century mixer must open simultaneously. T F
10. The Vialle vapor purge system removes _____ from the _____.
11. The vacuum developed inside the Vialle vaporizer operates the secondary diaphragm and valve. T F
12. An electric gasoline fuelock is used with Petrosystems conversion equipment. T F
13. Liquid propane flow is turned on and off by a _____ _____ in the Petrosystems vaporizer.
14. The flow volume of propane vapor is controlled by _____ _____ in a Petrosystems vaporizer.

CHAPTER 4

Propane Conversion Installations

Straight Propane Conversions

MOTOR FUEL TANKS Propane fuel tanks are constructed from heavy gauge steel. The external fuel tank fittings are shown in Figure 4-1.

As Figure 4-2 illustrates, the filler valve contains a double flow-check valve that allows fuel to flow into the fuel tank and prevents propane from flowing out of the valve.

An excess flow valve is located in the tank shutoff valve, as pictured in Figure 4-3. If the liquid fuel line was broken and the shutoff valve turned on, the excess flow valve would snap closed when fuel discharge exceeded the normal flow.

The relief valve will open if tank pressure becomes excessive. On internally mounted fuel tanks the relief valve must have a vent hose connected from the valve to the outside of the vehicle. Relief valve discharge pressure is the same as the tank pressure rating, 250 psi (1750 kp) on externally mounted tanks, and 312 psi (2,184 kp) on internally mounted tanks. Motor fuel tanks should only be filled to the 80 percent level. The fixed liquid level gauge contains a threaded plug with a number 54 drill-size orifice. When liquid discharge from the fixed liquid level gauge appears, filling of the fuel tank should be stopped. The space at the top of the tank allows for fuel expansion in hot weather. The most likely cause of relief valve discharge is overfilling of the fuel tank. A relief valve is pictured in Figure 4-4.

The magnetic float gauge contains a magnetic disc attached to the float arm. A similar magnetic disc is attached to the gauge pointer.

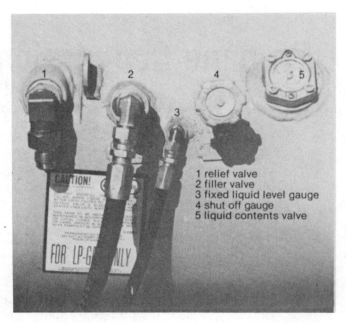

FIGURE 4-1. Propane Fuel Tank Fittings. *(Courtesy of Ontario Ministry of Transportation and Communication)*

Movement of the float attracts the pointer magnetic disc and indicates the fuel level. The pointer and magnetic disc may be replaced with a variable resistor if an electric fuel gauge is used. The variable resistor may be connected to the existing dash gauge, or a new under-dash gauge may be installed. On dual fuel applications, a switch may be installed to connect the dash gauge to either fuel tank sender unit. The magnetic float gauge is shown in Figure 4-5.

LIQUID FUEL LINES Fuel lines carrying liquid propane from the tank to the vaporizer must be stainless steel braided hose rated at 350 psi (2,450 kp). If the shutoff valve and the propane fuelock are closed, the fuel line could rupture because of the excessive pressure from the buildup of heat. A hydrostatic relief valve, as pictured in Figure 4-6, must therefore be installed in the liquid fuel line. Hydrostatic relief valve discharge pressure is 375–500 psi (2,625–3,500 kp). Regulations in most areas require the hydrostatic relief valve to be located at the bulkhead fitting where the fuel line comes through the vehicle floor, as illustrated in Figure 4-7.

Propane Conversion Installations 69

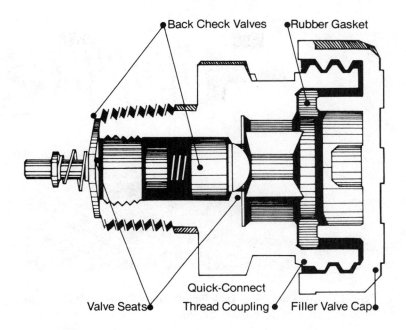

FIGURE 4-2. Fuel Tank Filler Valve. *(Courtesy of Ontario Ministry of Transportation and Communication)*

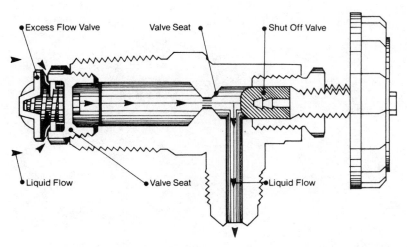

FIGURE 4-3. Fuel Tank Excess Flow Valve. *(Courtesy of Ontario Ministry of Transportation and Communication)*

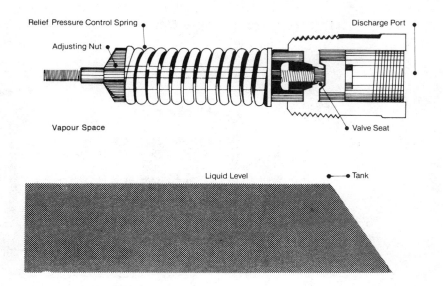

FIGURE 4-4. Fuel Tank Relief Valve. *(Courtesy of Ontario Ministry of Transportation and Communication)*

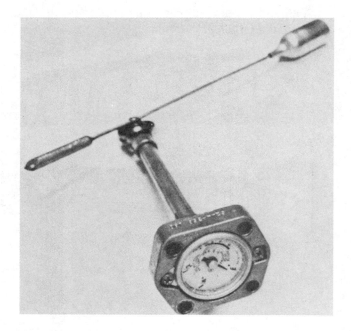

FIGURE 4-5. Fuel Tank Magnetic Float. *(Courtesy of Ontario Ministry of Transportation and Communication)*

Propane Conversion Installations

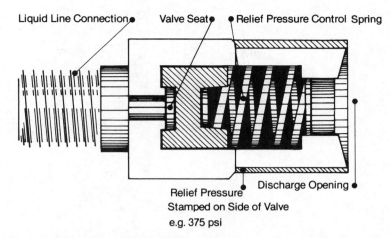

FIGURE 4-6. Fuel Line Hydrostatic Relief Valve. *(Courtesy of Ontario Ministry of Transportation and Communication)*

The numbered components in the straight propane system in Figure 4-7 are:

1. motor fuel tank
2. bulkhead fitting with hydrostatic relief valve
3. vacuum fuelock
4. vaporizer
5. air cleaner
6. mixer, carburetor

The lettered parts in Figure 4-7 are:

A. filler valve
B. vapor outlet (not used in some applications)
C. 80 percent fixed liquid level gauge
E. relief valve
F. relief valve discharge hose
G. shutoff valve
H. liquid propane hose
L. vacuum hose
M. heater hoses

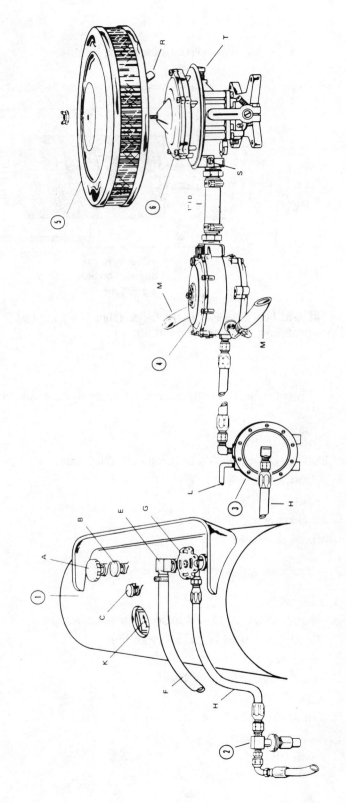

FIGURE 4-7. Straight Propane System. *(Courtesy of Impco Carburetion, Inc.)*

R. PCV hose fitting
S. vapor inlet
T. idle mixture adjustment

VAPORIZER INSTALLATION The vaporizer should be mounted close to the mixer. A short vapor hose will reduce the amount of priming required to start a cold engine. Many vaporizers are intended to be mounted vertically, with the vapor outlet facing downward. Vertical mounting allows oil in the propane to be discharged with the propane vapor going to the mixer. The disadvantage of vertical vaporizer mounting is that secondary diaphragm movement will be abnormal when severe car inclinations occur. Some vaporizers have the vapor outlet positioned near the top of the unit. The Vialle vapor purge system described in Chapter 3 eliminates oil buildup in this type of vaporizer. An accumulation of oil in the vaporizer can affect diaphragm movement and the air-fuel mixture. Without the vapor purge system, periodic vaporizer draining may be necessary for top-mounted vapor outlets.

The vaporizer should be mounted lower than the top of the radiator to ensure that the coolant remains at the proper level in the unit at all times. If the vaporizer must be located higher than the top of the radiator, the overflow container should be located at the same level as the vaporizer. Heater hoses must be connected to the vaporizer to allow the coolant to circulate through the unit. In cold climates adequate block heaters must be used to maintain vaporizer temperature during cold shutdown time. The vaporizer may not convert liquid propane to a vapor if the vaporizer temperature is below −25°F (−21°C). The most common method of connecting vaporizer heater hoses is illustrated in Figure 4-8. The vaporizer coolant flow is connected parallel to the heater core by means of Y-type fittings. The heater shutoff valve would be located between the Y fitting and the heater core. The point of the Y fittings must be positioned toward the engine.

An H-type fitting may be used to connect the vaporizer coolant flow parallel to the heater core, as shown in Figure 4-9. The vaporizer coolant flow may be in series with the heater core if a heater coolant control valve is not used.

MIXER INSTALLATION In a straight propane conversion the mixer may be adapted to the base of the gasoline carburetor. As noted in Chapter 3, some straight propane applications install the mixer on top of the gasoline carburetor. In straight propane installations the gasoline fuel pump is removed and a plate bolted to the pump opening in the block. When the

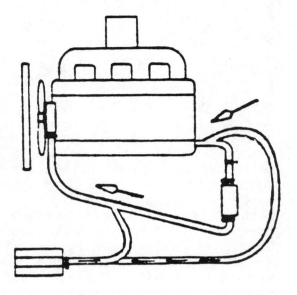

FIGURE 4-8. Vaporizer Y-type Heater Hose Connections. *(Courtesy of Vialle Autogas Systems)*

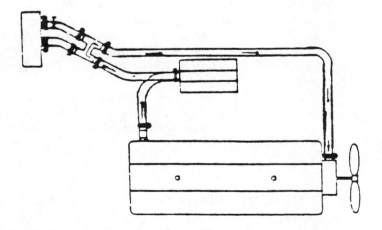

FIGURE 4-9. Vaporizer H-type Heater Hose Connection. *(Courtesy of Vialle Autogas Systems)*

gasoline carburetor base is used, the throttle shaft must be in good condition to prevent air leaks. All mixer gaskets and gasket surfaces should be carefully checked to eliminate air leaks. Rough idle operation and hard starting can be caused by air leaks into the mixer or intake manifold.

Dual Fuel Propane Conversions

DUAL FUEL AND STRAIGHT PROPANE SYSTEM VARIATIONS The same basic equipment is used for dual fuel and straight propane conversions. In dual fuel installations the gasoline fuel system is still operative, and fuelocks are installed in the gasoline and propane fuel lines. A dash switch allows the operator to select either fuel. The numbered components in the dual fuel system illustrated in Figure 4-10 are:

1. motor fuel tank
2. bulkhead fitting with hydrostatic relief valve
3. vacuum fuelock
4. vaporizer
5. mixer idle mixture adjustment
6. adapter
7. gasoline carburetor
8. gasoline fuelock

The lettered parts in Figure 4-10 are:

A. filler valve
B. vapor outlet (not used in most applications)
C. 80 percent fixed liquid level gauge
E. relief valve
G. shutoff valve
H. liquid line
I. EGR hose
J. EGR hose
K. vacuum hose to vacuum fuelock
L. heater hoses

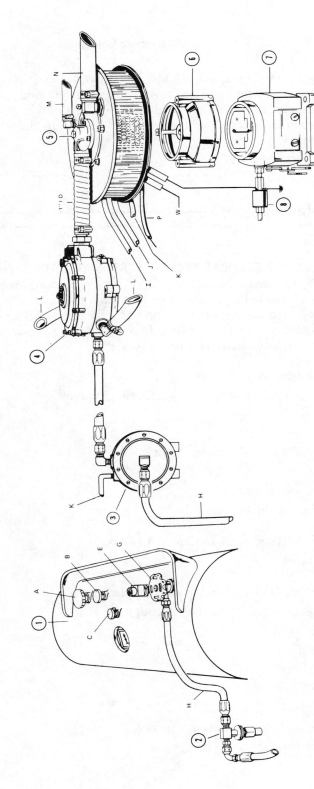

FIGURE 4-10. Dual Fuel System. *(Courtesy of Impco Carburetion, Inc.)*

M. gasoline vapor canister hose
N. PCV system hose
P. bowden cable
W. 12 V supply to gasoline fuelock switch

The propane mixer illustrated in Figure 4-11 is an Impco model 300 mixer, which was described in the previous chapter. In the gasoline mode, the bowden cable pulls the mixer open and thus allows unrestricted air flow to the gasoline carburetor. The vacuum to the propane fuelock is switched on and off by a vacuum switch under the mixer. An electric switch is used to open and close the circuit to the gasoline fuelock. Operation of both switches is controlled by a cam ring and bowden cable, as pictured in Figure 4-11. Most dual fuel systems use electric propane and gasoline fuelocks with a dash-mounted selector switch. The EGR vacuum hoses shown in Figure 4-10 are connected to an air cleaner temperature switch that is used to close the EGR vacuum circuit during engine warmup. The temperature switch was transferred from the gasoline air cleaner.

FUELOCKS A combination propane fuelock and filter is illustrated in Figure 4-12. Liquid propane enters at inlet 1 and passes through connector 2, depositing any heavy dirt particles inside the filter body. The fuel passes through bronze filter 4 to the upper plunger section of the valve. Spring 5 holds plunger 6 in a closed position when coil 7 is not energized. Plunger 6 is lifted when coil 7 is energized and allows fuel to flow through outlet 8 to the vaporizer. The filter can be serviced by removing nut 9.

The gasoline fuelock is installed in the gasoline line between the fuel pump and the carburetor. Figure 4-13 shows a gasoline fuelock. Fuel enters at inlet 1 and travels to the upper seat section of the valve. Spring 2 holds plunger 3 in a closed position when coil 4 is not energized. Plunger 3 is lifted when coil 4 is energized, and fuel is able to flow through outlet 5 to the carburetor. If an electrical defect should occur, manual control valve 6 may be opened to allow fuel flow through the fuelock. Inlet and outlet fittings are clearly identified and must never be reversed.

Regulations in some areas require the installation of a vacuum switch in series with the electric circuit to the propane fuelock. The vacuum switch will close when the engine is cranked to supply power to the propane fuelock. If the engine stalls and the ignition switch is left on, the vacuum switch opens the circuit to the fuelock. Figure 4-14 shows a typical vacuum switch.

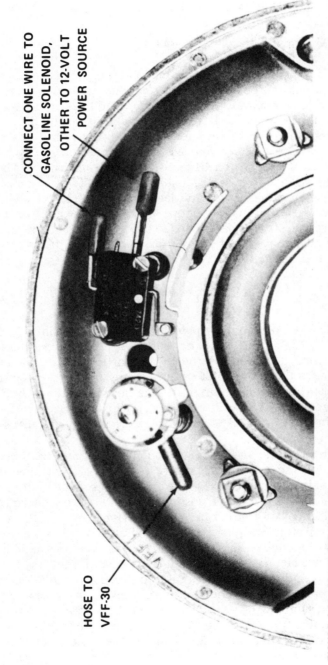

FIGURE 4-11. Model 300 Mixer Control Switches. *(Courtesy of Impco Carburetion, Inc.)*

Propane Conversion Installations

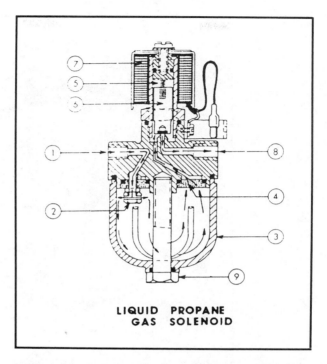

FIGURE 4-12. Propane Fuelock. *(Courtesy of Alternative Fuel Systems, Ltd.)*

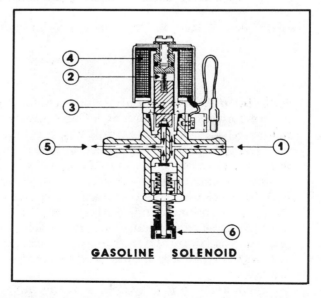

FIGURE 4-13. Gasoline Fuelock. *(Courtesy of Alternative Fuel Systems, Ltd.)*

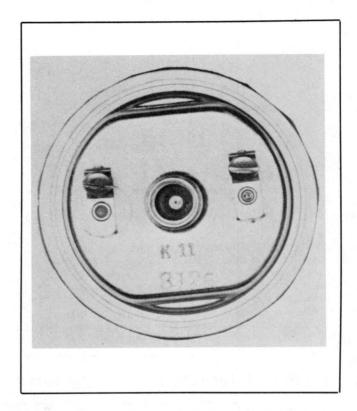

FIGURE 4-14. Vacuum Switch. *(Courtesy of Alternative Fuel Systems, Ltd.)*

DUAL FUEL WIRING Electrical circuits vary considerably, depending on the type of conversion equipment. A Garretson wiring diagram is illustrated in Figure 4-15. Two double-throw toggle switches are dash mounted. The switches are used for fuel selection and priming. A vacuum switch is located in the electric circuit to the propane fuelock. The vacuum switch contains a set of normally closed contacts between the bottom terminal and the top terminal. A set of normally open contacts is located between the center terminal and the top terminal. The 12 V supply wire connected to the fuse block in Figure 4-15 must have power available when the ignition switch is on and the engine is being cranked. The wire connected to the starting motor should be connected to the main starting motor terminal.

When the prime switch is in the auto prime mode, contact is completed between the bottom and center terminals in the prime switch.

Placing the fuel selector switch in the LPG position completes the circuit between the lower and center switch terminals. When the engine is being cranked, current will flow from the starting motor main terminal through the lower contacts on both switches. Current will flow from the LPG switch to the prime solenoid, and through the closed vacuum switch contacts to the LPG fuelock. Once the engine starts, the normally open vacuum switch contacts will close. Current now flows from the 12 V supply wire through the lower LPG switch contacts and vacuum switch contacts to the LPG fuelock.

Selection of the manual prime mode closes the contacts between the upper and center prime switch terminals. Current will flow from the LPG switch lower contacts through the prime switch upper contacts to the prime solenoid and the propane fuelock. The prime switch is spring loaded in the manual prime position. The manual prime mode may have to be used for a few seconds before starting a cold engine.

When the gasoline mode is selected, electrical contact is completed between the center terminal and upper terminal on the fuel selector switch. The gasoline fuelock will be energized from the upper fuel selector switch contacts. When changing from the gasoline mode to the

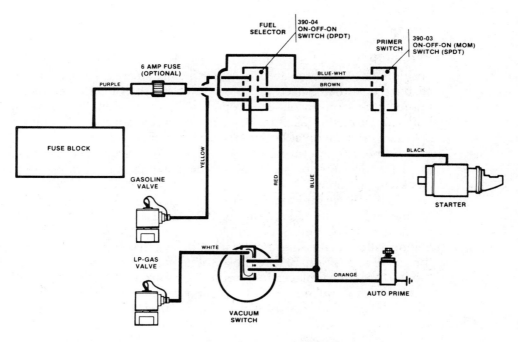

FIGURE 4-15. Dual Fuel Wiring. *(Courtesy of Garretson Equipment Company, Inc.)*

FIGURE 4-16. Rotary Fuel Selector Switch. *(Courtesy of Alternative Fuel Systems, Ltd.)*

LPG mode, the fuel selector switch must be placed in the center neutral position for a brief time until the engine begins stalling. The gasoline in the carburetor must be used up before selecting the LPG mode. Switching immediately from gasoline to propane will cause the engine to operate on both fuels until the gasoline is used out of the carburetor. The engine will operate very roughly if both fuels are burned.

Some manufacturers use a four-position rotary fuel selector switch, as pictured in Figure 4-16. The selector switch is always rotated in a clockwise direction.

In the fill carburetor (FC) position, both fuelocks are energized. The engine will continue to operate on LPG until the gasoline carburetor is filled. If gasoline enters the intake manifold with propane vapor, the engine will begin missing and the selector switch will have to be rotated to the gasoline position. Before the gasoline position is changed to the LPG mode, the switch is placed in the empty carburetor (EC) position until the engine begins stalling. The EC position deenergizes both fuelocks and allows the gasoline to be used up in the carburetor before LPG is selected.

Generally Accepted Regulations

MOTOR FUEL TANKS Regulations covering the installation of propane conversion equipment are a state or provincial responsibility. Reference should always be made to local regulations. We will discuss some of the more widely accepted regulations. Inspection of motor fuel tanks is usually the responsibility of the authority in charge of high-pressure vessels. In most countries, approval is necessary for importing or manufacturing propane fuel tanks. Each tank must also be inspected and stamped by the local authority in charge of high-pressure vessels. Listed below are some common regulations regarding motor fuel tanks.

1. New fuel tanks should be purged four times with propane vapor at 15 psi (105 kp) prior to initial filling.
2. In straight propane conversions, the gasoline fuel tank must be removed.
3. Externally mounted fuel tanks are rated at 250 psi (1,750 kp), and internally mounted fuel tanks are rated at 312 psi (2,184 kp).
4. Internally mounted tanks must have remote hoses extending outside the vehicle on the filler valve, an 80 percent fixed liquid level outlet, and a relief valve. The fixed liquid level outlet may be referred to as an outage valve. Figure 4-17 pictures a remote vent hose. State regulations should be checked for the type of vent hose required. Local regulations may specify an upward or a downward direction for the vent hose in relation to the fuel tank.
5. An alternative to using remote hoses is to completely seal the passenger compartment and all electrical devices from the area where the tank is mounted.
6. Trunk venting at the lowest point is necessary for trunk-mounted tanks.
7. Internally mounted tanks require an 80 percent stop fill valve in the filler inlet of the tank. This type of valve can be recognized by a four-bolt flange on the tank fill valve. State regulations in some areas may not require an 80 percent stop fill valve on internally mounted tanks.
8. Tanks must be attached to the vehicle structure with 1/2-in (12-mm) bolts. If the vehicle structure is less than 1/8 in (3 mm) thick in the tank bolt area, metal reinforcing plates 16 in^2 (102 cm^2) in area and 1/8 in (3

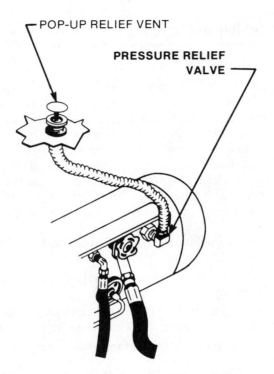

FIGURE 4-17. Propane Tank Vent Hose. *(Courtesy of Petrosystems International, Inc.)*

mm) thick must be used. For supported tanks the reinforcing plates are placed under the vehicle floor, and for suspended tanks the plates are placed above the floor.

9. Tank mounting requirements.
   a. The tank should be mounted in such a position that damage will be minimal in case of a collision.
   b. Tank heat shielding is required if the tank is located closer than 8 in (20 cm) to the exhaust system or engine.
   c. Minimum clearance between the fuel tank and the exhaust system or engine is 4 in (10 cm).
   d. Roof mounting of fuel tanks is prohibited.
   e. The tank shall not project beyond the sides of the vehicle.
   f. Tank mounting forward of the front axle is not allowed.
   g. The fuel tank shall not extend beyond the rear bumper.

Propane Conversion Installations

h. The structural part of the vehicle must be 6 in (15 cm) above the top of the fuel tank.
i. All valves must be in a protected position.
j. Access to all valves is essential.
k. The magnetic float gauge, if used, must be in view of the person refueling.
l. External fuel tanks must be positioned 1 in (2.5 cm) above the lowest nonrotating part of the vehicle, excluding brake components. (NOTE: Road clearance requirements vary in different locations.)
m. Tanks located in the passenger compartment must be in a separate sealed enclosure, and the enclosure vented outside the vehicle.

The remote fill hose and fixed liquid level hose are shown in Figure 4-18. Remote hoses may terminate in a fill box, as illustrated in Figure 4-19. On some vehicles it may be convenient to attach the remote hoses to the gasoline filler neck, as indicated in Figure 4-20.

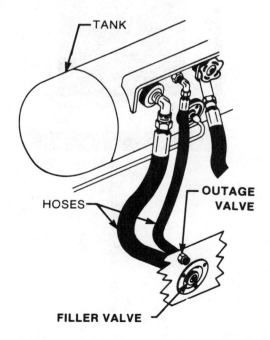

FIGURE 4-18. Propane Tank Remote Hoses. *(Courtesy of Petrosystems International, Inc.)*

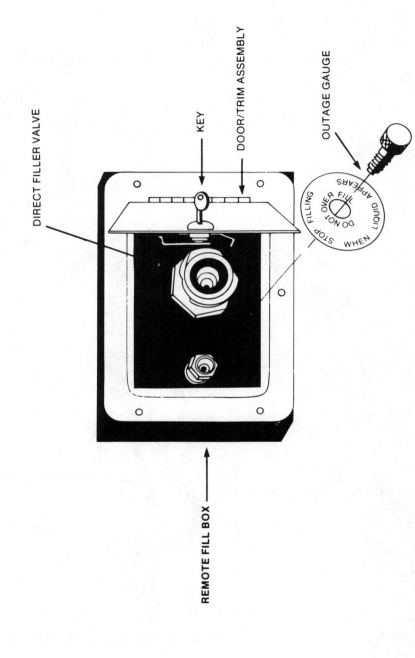

FIGURE 4-19. Remote Fill Box. *(Courtesy of Petrosystems International, Inc.)*

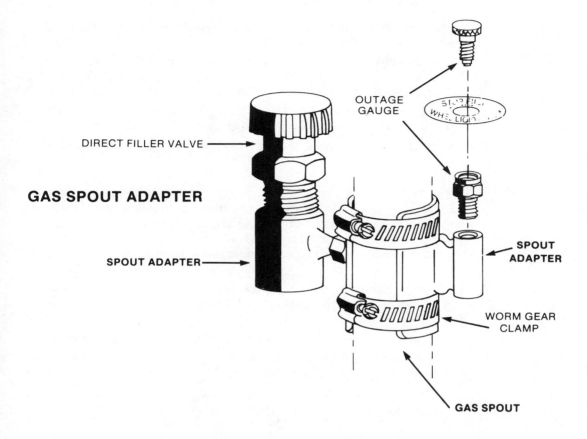

FIGURE 4-20. Remote Hoses Attached to Gasoline Tank Filler Neck. *(Courtesy of Petrosystems International, Inc.)*

HOSE REQUIREMENTS The following items are some of the most important regulations regarding propane fuel hose installation.

1. All hoses used for liquid line, remote fill line, 80 percent fixed liquid level line, and relief vent line must be stainless steel braid propane fuel line rated at 350 psi (2,450 kp).
2. Minimum inside hose diameters are as follows:
   a. Liquid line 5/16 in (8 mm)
   b. 80 percent fixed liquid level line 1/4 in (6 mm)
   c. Fill line 5/8 in (16 mm)
3. All hoses, except the relief valve hose, must be attached to the vehicle structure at maximum intervals of 2 ft (61 cm) by means of plastic-

coated metal clamps. Some areas may allow nylon ties to be used to attach the fuel hoses.
4. The liquid fuel line must be equipped with a hydrostatic relief valve located at the bulkhead fitting. The relief valve is rated at 375-500 psi (2,625-3,500 kp).
5. Fuel hoses must have a bulkhead fitting, or must be protected with a grommet where they extend through the vehicle structure.
6. All male fittings must be coated with pipe dope before installation.

SAFETY PRECAUTIONS Propane fuel systems have many built-in safety features. Vehicles fueled with propane have an excellent safety record. Propane equipment must have approval from such organizations as the Underwriters Laboratory. Propane fuel systems will continue to enjoy satisfactory safety records if a few basic safety regulations are observed:

1. Before disconnecting a liquid propane line, close the tank shutoff valve and operate the engine until it stops.
2. Liquid propane expands 270 times when it changes to a vapor. Liquid propane, or propane vapor, should not be released in a closed shop.
3. If propane is released in a closed area, extinguish all sources of ignition such as pilot lights, and ventilate the area.
4. Leak testing must be done on completion of all propane conversions. Liquid leak-detecting solutions or an electronic tester may be used. The propane fuelock must be energized when testing for leaks.
5. When propane odor occurs in a vehicle, all fittings should be leak-tested. Some odor is usually emitted from the vapor hose and air cleaner.
6. Periodically all fuel hoses should be visually inspected and all fittings should be leak-tested.
7. Propane-fueled vehicles should not be stored near pits or sewers in a shop. Propane vapor is heavier than air and settles into low areas.
8. Keep welding equipment and other sources of ignition away from propane vehicle service areas.
9. Some local regulations require tank shutoff valves to be closed when propane-fueled vehicles are stored indoors.
10. Do not smoke while servicing or refueling propane-equipped vehicles.
11. Obey all regional conversion and fire regulations.

Questions

1. If the liquid propane fuel line is broken and the tank shutoff valve is on, propane will escape until the tank is empty.   T   F
2. The hydrostatic relief valve on the liquid propane line should be located under the vehicle hood.   T   F
3. The vaporizer should be mounted lower than the top of the _____ _____.
4. Vaporizer temperature must be kept above _____ degrees F for cold weather starting.
5. The heater coolant control valve should control coolant flow through the vaporizer.   T   F
6. A dual fuel system may be switched instantly from gasoline to propane.   T   F
7. When the rotary fuel selector switch is in the FC position, both fuelocks are energized.   T   F
8. Liquid propane fuel hose must be attached to the vehicle structure at maximum intervals of _____.
9. Liquid propane expands _____ times when it changes to a vapor.
10. The minimum inside diameter of remote fill lines is _____.

CHAPTER 5

Engine Tuning for Propane Fuel

Ignition Basics

IGNITION OPERATION The primary ignition system consists of the battery, ignition switch, resistor, primary coil winding, and electronic module. Many late model ignition systems have eliminated the primary circuit resistor. The secondary coil winding, distributor cap, rotor high-tension wires, and spark plugs make up the secondary ignition system. Distributor pickup coil leads are connected from the distributor to the electronic module, as indicated in Figure 5-1.

A reluctor, which is fastened to the distributor shaft, is designed with a high point for each cylinder of the engine. The reluctor may also be referred to as a timer core or an armature, depending on the manufacturer. Reluctor high points rotate past the head on the pickup coil. A permanent magnet is located in the pickup coil assembly.

When the reluctor high points are out of alignment with the pickup coil, as shown in Figure 5-2, the high points will not affect the magnetic strength of the pickup coil. Under this condition, the electronic module allows current to flow through the primary circuit. Primary current flow creates magnetic buildup in the coil.

When the reluctor high point is moved into alignment with the pickup coil, magnetic field of the pickup coil is strengthened, as indicated in Figure 5-3. When the reluctor high point begins moving out of alignment with the pickup coil, the magnetic field collapses across the pickup winding. The resulting induced voltage in the pickup coil signals the electronic module to open the primary circuit. As the primary circuit

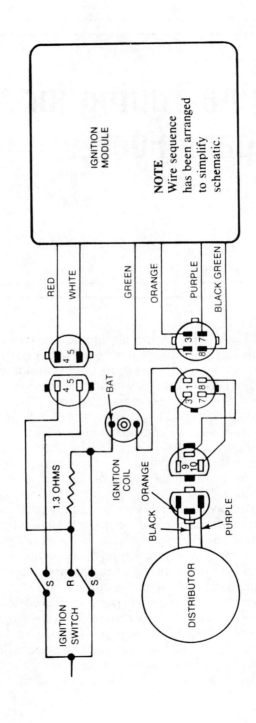

FIGURE 5-1. Ignition System Diagram. *(Courtesy of Ford Motor Company)*

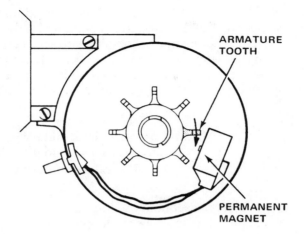

FIGURE 5-2. Reluctor High Points out of Alignment. *(Courtesy of Ford Motor Company)*

opens, a sudden magnetic collapse occurs across the ignition coil windings, and it induces a high voltage in the secondary winding. The high secondary voltage causes a small amount of current to flow across the spark plug gap and ignite the air-fuel mixture.

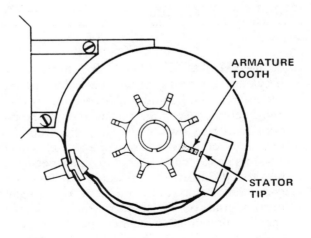

FIGURE 5-3. Reluctor High Point in Alignment. *(Courtesy of Ford Motor Company)*

SECONDARY VOLTAGE REQUIREMENTS Maximum secondary coil voltage varies from 22 kv to 35 kv, depending on the type of electronic ignition system. The normal secondary voltage required to fire the spark plug would be approximately 10 kv (kv is the abbreviation for kilovolts, or thousands of volts). Reserve secondary voltage is the difference between normal required voltage and maximum available voltage, as indicated below:

Maximum available secondary voltage	25,000
Normal required secondary voltage	10,000
Reserve secondary voltage	15,000

Reserve secondary voltage is necessary to compensate for the high resistance that develops at the spark plug electrodes as a result of gradual erosion. When spark plug gaps become wider, the normal required secondary voltage must be increased. Under heavy engine load conditions, higher cylinder pressure increases secondary voltage requirements, and reserve secondary voltage becomes essential. Ignition misfiring during hard acceleration often occurs because secondary reserve voltage is lacking. High resistance in the secondary circuit increases the normal required voltage and reduces the voltage reserve. A defective coil or low primary current will lower the maximum secondary voltage and reduce the reserve voltage. Leakage in the distributor cap will also lower the maximum secondary voltage.

Propane Ignition Requirements

SECONDARY IGNITION CIRCUIT In contrast to gasoline, propane vapor between the spark plug electrodes increases the normal required secondary voltage under all operating conditions, as indicated in Figure 5-4. Secondary voltage requirements are particularly high on sudden acceleration at low speeds. The general condition of the ignition system is of greater concern when propane fuel is used. Many electronic ignition systems have a wide spark plug gap specification of 0.045 to 0.080 (1.143 mm to 2.032 mm). On a propane-fueled engine, spark plugs with a wide gap specification should be set at 0.035 (0.889 mm). A gap of 0.030 (0.762 mm) would be satisfactory on spark plugs with an original gap specification of 0.035 (0.889 mm). When gasoline fuel is used, spark plug gaps are usually 0.045 to 0.080. All spark plugs with these wide gap specifications should be set to 0.035 on a propane-fueled engine.

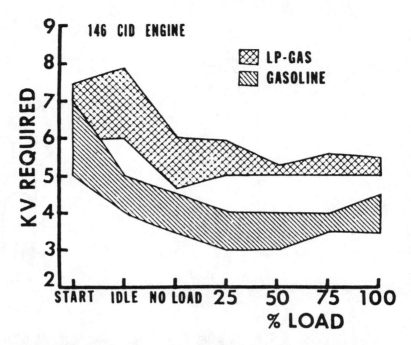

FIGURE 5-4. Propane Secondary Ignition Voltage Requirements. *(Courtesy of Champion Spark Plug Company of Canada, Ltd.)*

The spark plug heat range indicates the ability of a spark plug to conduct heat. Figure 5-5 illustrates the difference between a higher heat-range spark plug, and a spark plug with a lower heat range.

The center spark plug electrode becomes much hotter than the ground electrode. Heat must be dissipated from the center electrode through the insulator and shell of the spark plug to the cooling system. In a higher heat-range spark plug, the distance from the electrode tip to the point of contact between the insulator and the shell is much longer than in a lower heat-range plug. The electrodes remain hotter in a high heat-range spark plug because of the longer heat path through the plug. Propane-fueled engines operating under light duty conditions should have the same spark plug as a gasoline-fueled engine. When a heavy duty application is converted from gasoline to propane, spark plugs one range colder than originally specified should be installed because propane vapor entering the combustion chamber has no cooling effect. Furthermore, the ignition temperature of propane is higher than gasoline.

Normal spark plug carbon conditions in a propane-fueled engine are indicated by a small amount of white carbon on the plug insulator. Burned spark plug electrodes can be caused by incorrect air-fuel

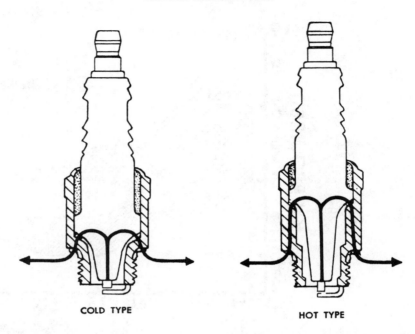

FIGURE 5-5. Spark Plug Heat Range. *(Courtesy of Champion Spark Plug Company of Canada, Ltd.)*

mixture, inaccurate ignition timing or distributor advance curves, or high heat-range spark plugs. Rich gasoline air-fuel ratios will cause black carbon to build up on spark plugs, excessively rich propane air-fuel mixtures will not cause such deposits.

When spark plug operating temperatures are lowered, voltage requirements increase, as shown in Figure 5-6. Increasing the normal required voltage lowers the secondary reserve voltage. Propane-fueled engines with excessively low heat-range spark plugs may misfire on hard acceleration owing to reduced secondary reserve voltage.

Resistance and leakage defects in secondary ignition circuits may be more noticeable on a propane-fueled engine owing to the higher required secondary voltage. The maximum resistance in 8 mm spark plug wires should be 12,000 ohms per ft. Resistance in 7 mm plug wires should not exceed 8,000 ohms per ft. Spark plug wires that fire one after the other in the firing order should be separated in the secondary wiring harness, as indicated in Figure 5-7.

The secondary voltage required to move current from the hot center electrode to the cooler ground electrode is lower than the voltage

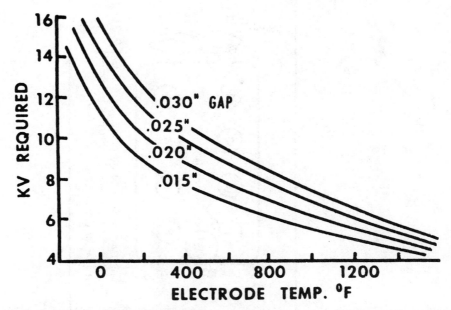

FIGURE 5-6. Effect of Low Heat Range Spark Plugs. *(Courtesy of Champion Spark Plug Company of Canada, Ltd.)*

required to move current through the electrodes in the reverse direction, as shown in Figure 5-8. Additional heat on the center spark plug electrodes creates a thermocouple action at the electrodes and, as a result, the electricity tends to move from a hot surface to a cooler surface.

The primary coil terminals must be connected to the battery with correct polarity. In negative ground battery systems, the negative primary coil terminal is connected to the ignition points, or to the electronic module in an electronic system. Correct primary connections ensure that the primary current flow and magnetic buildup will be in the desired direction. The result of proper magnetic buildup in the coil is negative polarity at the center spark plug electrodes, and current flow from the center electrode to the ground electrode, as indicated in Figure 5-9. Incorrect primary coil connections result in reversed primary current flow, magnetic buildup, and secondary voltage and current. Reversed secondary polarity will increase the normal required secondary voltage and misfiring may occur on acceleration. Correct secondary coil polarity is more critical on a propane-fueled engine than on a gasoline-fueled engine.

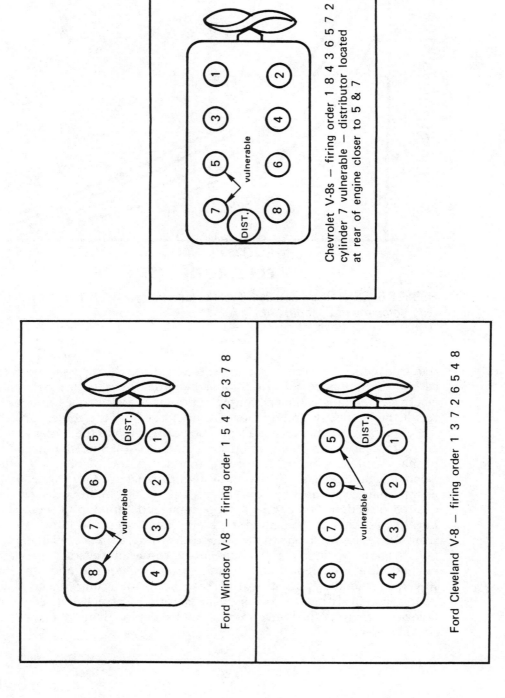

FIGURE 5-7. Spark Plug Wire Routing. *(Courtesy of Impco Carburetion, Inc.)*

Engine Tuning for Propane Fuel

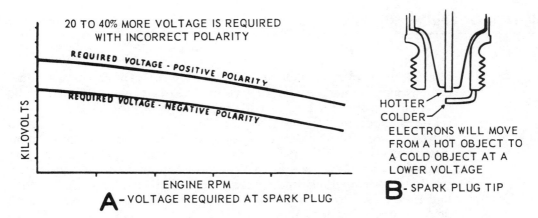

FIGURE 5-8. Secondary Polarity at Spark Plug Electrodes. *(Courtesy of Ignition Manufacturers Institute)*

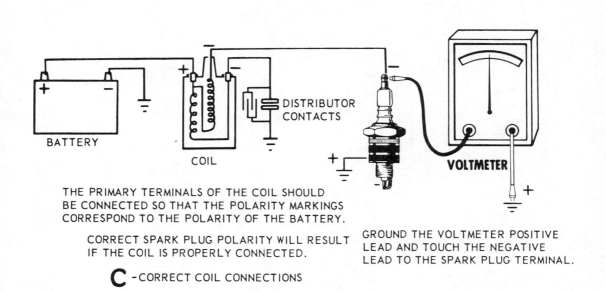

FIGURE 5-9. Primary Coil Connections and Secondary Coil Polarity. *(Courtesy of Ignition Manufacturers Institute)*

Distributor Timing and Advances

PURPOSE OF DISTRIBUTOR ADVANCE The average basic timing on a gasoline-fueled engine would be 8 degrees before top dead center (BTDC) at idle speed. Piston speed increases with engine rpm, but the air-fuel mixture requires approximately the same burning time. When the piston speed increases, the timing must be advanced to maintain maximum combustion pressure on the piston. Failure to advance the spark timing in relation to engine rpm would result in a loss of power, because the piston would be moving down in the power stroke before the air-fuel mixture had time to start burning. Ignition spark timing is advanced from 12 degrees at 1,200 rpm to 40 degrees at 3,600 rpm, as illustrated in Figure 5-10. Effective combustion ends at 23 degrees after top dead center (ATDC) on the power stroke, as shown in Figure 5-10.

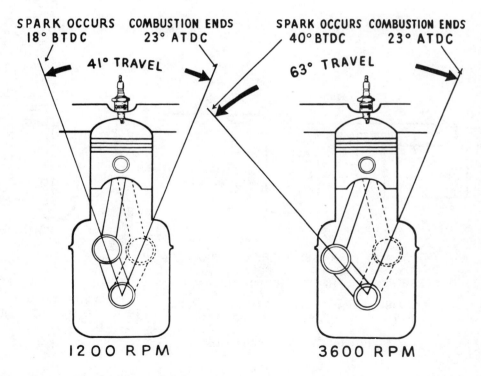

FIGURE 5-10. Ignition Spark Advance. *(Courtesy of Ignition Manufacturers Institute)*

CENTRIFUGAL ADVANCE OPERATION The centrifugal advance mechanism contains pivoted advance weights that rotate with the distributor shaft. Outward movement of the weights is proportional to engine and distributor shaft speed. Calibrated springs allow specific weight movement in relation to engine rpm, as pictured in Figure 5-11.

The reluctor is rotated ahead of the distributor shaft by the weight action resulting in earlier spark timing at each spark plug. The centrifugal advance mechanism provides ignition spark advance in relation to engine speed. Insufficient spark advance would allow the piston to move down in the power stroke before maximum combustion pressure occurred, and the result would be a loss of engine power.

FIGURE 5-11. Centrifugal Advance Mechanism. *(Courtesy of General Motors of Canada, Ltd.)*

VACUUM ADVANCE OPERATION The vacuum advance mechanism contains a vacuum diaphragm linked to the pickup coil plate. A sealed diaphragm chamber is connected to the intake manifold. Some vacuum advance diaphragms are connected directly to the intake manifold, whereas others are connected to the ported manifold vacuum above the throttles. Lean air-fuel mixtures are slower burning than rich air-fuel mixtures. Additional spark advance provides improved economy with leaner, normal cruise air-fuel mixtures. Rich full load mixtures require a reduction in ignition spark advance to prevent detonation. A high manifold vacuum under normal cruise conditions moves the vacuum advance diaphragm. The vacuum advance provides additional spark advance by rotating the pickup coil in the opposite direction in relation to the distributor shaft rotation. The loss of manifold vacuum under heavy load conditions allows the vacuum diaphragm to retard the spark advance. Ignition spark advance in relation to engine load is controlled by the vacuum advance. Figure 5-12 shows a typical vacuum advance mechanism.

Propane Distributor Advance Requirements

INITIAL TIMING Propane fuel is slightly harder to ignite in the combustion chamber than gasoline. The initial timing should be advanced approximately 5 degrees on many engines converted from gasoline to propane. Initial timing on most propane-fueled engines should be set between 10 and 14 degrees. The timing specification of 16 to 18 degrees on some late model engines should be maintained at the original specifications for propane fuel. The emission laws of some states, however, require that original timing and distributor advance specifications be maintained.

DISTRIBUTOR ADVANCE SPECIFICATIONS A propane-fueled engine requires less spark advance at higher rpm than a gasoline-fueled engine, as illustrated in Figure 5-13. Initial timing plus maximum centrifugal advance should not exceed 30 crankshaft degrees. An initial timing setting of 12 degrees would require a maximum centrifugal advance of 18 degrees. The maximum vacuum advance should be 15 crankshaft degrees. Distributor advance calibrations are most critical on heavy duty applications where additional heat is encountered.

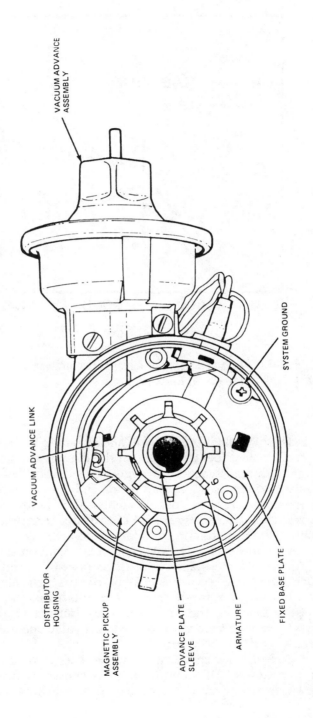

FIGURE 5-12. Vacuum Advance Mechanism. *(Courtesy of Ford Motor Company)*

104 Alternate Automotive Fuels

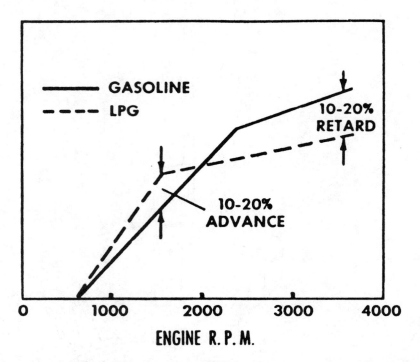

FIGURE 5-13. Propane Distributor Advance Requirements. *(Courtesy of Champion Spark Plug Company of Canada, Ltd.)*

EFFECTS OF INCORRECT IGNITION SPARK ADVANCE The ignition spark occurs before top dead center (TDC) on the compression stroke. Exact spark advance timing depends on engine speed and load. Cylinder pressure gradually increases as the piston moves upward, and maximum combustion pressure occurs when the piston is at TDC of the compression stroke. Figure 5-14 illustrates a normal cylinder pressure graph with the piston on the compression stroke.

A combustion pressure graph of late ignition spark timing is shown in Figure 5-15. Reduced combustion pressure occurs when the piston moves down in the power stroke before the air-fuel mixture has time to start burning. Engine power loss results from the lower combustion pressure. Because the air-fuel mixture burns late in the power stroke, excess heat is present when the exhaust valve opens. As a result, exhaust valve life will be shortened.

Early ignition spark timing causes excessive combustion before the piston reaches TDC compression, as shown in Figure 5-16. Extreme pressures occur in the cylinder as the piston moves upward and

Engine Tuning for Propane Fuel 105

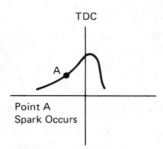

FIGURE 5-14. Normal Cylinder Pressure Graph.

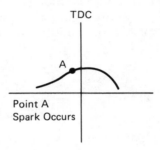

FIGURE 5-15. Combustion Pressure Graph Resulting from Late Ignition Timing.

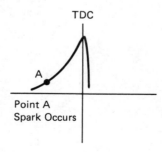

FIGURE 5-16. Combustion Pressure Graph Resulting from Early Ignition Timing.

compresses the expanding air-fuel mixture. Engine power is lost and excessive piston heat generated when the expanding air-fuel mixture attempts to drive the piston back down on the compression stroke. When the piston goes past TDC compression, the remaining air-fuel mixture may explode suddenly as a result of the excessive heat in the combustion chamber. Severe detonation and piston damage may result from the rapid explosion of the air-fuel mixture.

CALIBRATING DISTRIBUTOR ADVANCES FOR PROPANE-FUELED ENGINES Gann plates may be installed on the centrifugal advance mechanism to provide the correct advance for propane fuel. The movement of the advance mechanism is limited by the Gann plate, as shown in Figure 5-17. Various plates are available for each type of distributor to provide the desired amount of advance.

Electronic correction of distributor advance calibrations is possible with the dual curve ignition control unit shown in Figure 5-18.

The dual curve ignition control unit is designed for dual fuel engines. Distributor advance calibrations are altered for propane fuel, but remain

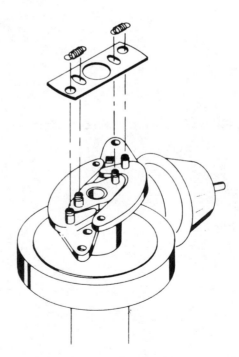

FIGURE 5-17. Gann Plate. *(Courtesy of Gann Products Company, Inc.)*

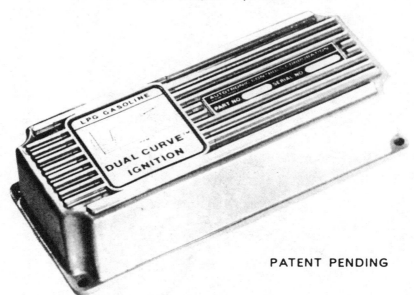

FIGURE 5-18. Dual Curve Ignition Control Unit. *(Courtesy of Autotronic Controls Corporation)*

unchanged in the gasoline mode. Wiring connections for the control unit are shown in Figure 5-19. The fuel mode signal to the control unit passes through the yellow wire connected to the gasoline fuelock.

Centrifugal advance mechanisms may be altered by partial welding of the advance slots. Many vacuum advance units have an adjusting nut located inside the vacuum outlet. Counterclockwise rotation of the adjusting nut, with an allen wrench, decreases the degrees of vacuum advance. Vacuum advance movement may be limited by partial welding of the advance arm slot. As mentioned earlier, emission laws in some states may forbid the alteration of distributor advances. Changes in initial timing or distributor advances made for propane fuel may cause detonation when gasoline is used.

Infrared Adjustment and Diagnosis of Propane Fuel Systems

EMISSION LEVELS The ideal gasoline air-fuel ratio is 14.5:1. For propane fuel, the desired air-fuel ratio is 15.5:1. With an air-fuel ratio of 15.5:1 the propane mixer supplies one pound of fuel for every 15.5 pounds of air

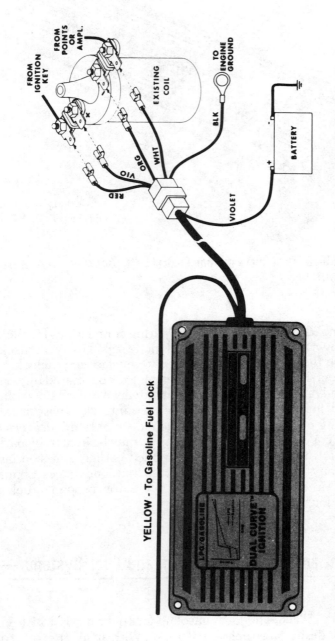

FIGURE 5-19. Dual Curve Ignition Control Unit Wiring Connections. *(Courtesy of Autotronic Controls Corporation)*

that rush into the engine. Carbon monoxide (CO) levels are directly proportional to air-fuel ratios. A gasoline air-fuel ratio of 14.1:1 provides a CO level of 1 percent. The ideal propane air-fuel ratio of 15.5:1 provides a CO level of approximately 1 percent, as shown in Table 5-1.

TABLE 5-1. Air-Fuel Ratios and Carbon Monoxide Levels

Natural Gas	–	18.0	*16.2*	14.8	14.3	13.8	–
Butane Propane	18.0	16.3	*15.4*	14.2	13.6	12.8	12.0
Gasoline	16.0	15.0	14.0	*13.0*	12.0	11.0	10.0
Carbon Monoxide %	0	1	2	3	4	5	–

In many states or provinces CO levels must be adjusted to meet emission regulations. In areas where CO levels are not dictated by emission laws, 1 percent CO would be acceptable at idle speed. At 2,000 engine rpm, CO levels should decrease from the reading at idle speed. Full load CO levels should be about 3 percent, which indicates an air-fuel ratio of 14.4:1. Actual full throttle conditions on the road, or on a dynamometer, are necessary to obtain accurate full load mixture settings. A portable 12-v-powered analyzer can be used in road testing. The infrared tester pictured in Figure 5-20 reads CO levels and unburned hydrocarbons in parts per million (ppm). On cars equipped with catalytic converters, the infrared pickup must be installed ahead of the converter to provide accurate readings. Most manufacturers recommend disconnecting air pumps when checking emission levels.

In many states hydrocarbon (HC) levels must be adjusted according to emission regulations. Excessively rich or lean air-fuel ratios will result in high HC levels because combustion is incomplete. Ignition or compression defects will cause cylinder misfiring and high HC levels. Where specific HC levels are not required by emission laws, 200 ppm would be acceptable at idle speed. Under cruise conditions HC levels should be less than 50 ppm.

Idle mixtures are adjusted by turning the idle mixture screw on the propane mixer. (See Chapter 3 for the location of the idle mixture screw on various mixers.) Cruise mixtures are determined by the taper on the gas valve in mixers using the tapered gas valve principle. On a Garretson system, cruise mixtures are adjusted by the venturi plate setting and the secondary valve spring adjustment. (See Chapter 3 for venturi plate settings.) Many other mixers using the venturi ring principle are not adjustable. Full throttle mixtures are adjusted with the full load screw in the mixer or vapor hose.

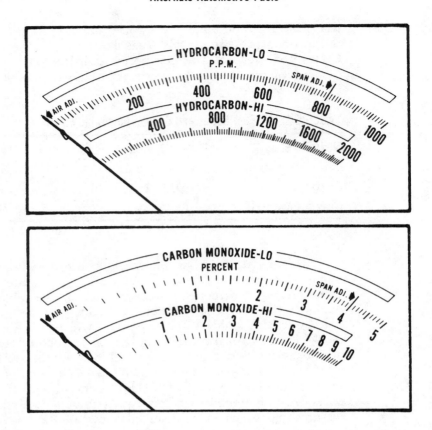

FIGURE 5-20. Infrared Tester. *(Courtesy of Applied Power, Ltd.)*

ADJUSTING IMPCO EC1-EQUIPPED VAPORIZERS The following steps should be taken when adjusting an EC1 device:

1. Operate the engine until coolant reaches normal operating temperature.
2. Disconnect and plug EC1 hoses.
3. Operate engine at 2,000 rpm.
4. Reconnect EC1 hoses; engine should slow down 100 rpm and CO levels should be 0.1-0.2 percent. Bleed screw A in Figure 5-21 can be rotated to adjust the EC1 speed change.
5. With EC1 hoses connected, slowly back out engine speed screw. EC1 should go out of action at 1,000-1,200 rpm, as evidenced by an increase

in engine rpm. Transition speed is adjusted by removing plug B in Figure 5-21 and turning the adjusting screw. Clockwise rotation of the adjusting screw increases the transition speed.

Diagnosis of Propane-Fueled Engines

HARD STARTING

1. Check cranking speed; slow cranking speed could be caused by a defective battery, high resistance in the battery cables, or a defective starting motor.
2. Test for spark at plug wires when cranking the engine. If no spark occurs, repair ignition system.
3. Be sure there is 9 v or over at the positive primary coil terminal when cranking the engine. Low voltage at the coil primary could be caused by a defective primary circuit resistor bypass circuit on ignition systems with a primary circuit resistor.
4. Energize propane fuelock and priming device and check for propane vapor through the vapor hose or prime solenoid. Eliminate all sources of ignition when making this test. If fuel vapor is not present, check filter and fuel supply from the tank. Observe safety precautions mentioned in Chapter 4 when servicing filter or liquid fuel line.
5. Check for air leaks into the intake manifold. Lean air-fuel mixtures can result in hard starting with gaseous fuels.
6. Test priming circuit; hard starting can result from overpriming or underpriming.

ROUGH IDLE OPERATION

1. Check cylinder compression.
2. Check spark plugs, plug wires, and distributor cap.
3. Check for vacuum leaks into intake manifold or propane mixer.
4. Test exhaust gas recirculation (EGR) valve and control devices. Rough idling at low speed could result from an open EGR valve.

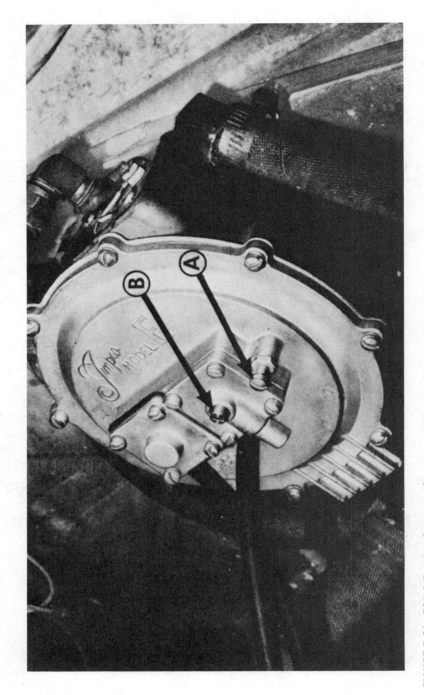

FIGURE 5-21. EC1 Adjusting Screws. *(Courtesy of Impco Carburetion, Inc.)*

EXCESSIVE POWER LOSS

1. Check cylinder compression.
2. Check ignition timing and distributor advances.
3. Check air cleaner for restriction.
4. Check exhaust system for restriction.
5. An enclosed air cleaner will improve power output.
6. Be sure mixer cfm rating and vaporizer hp rating are sufficient for the engine.
7. Test CO levels; lean mixtures could result from plugged fuel filter, defective vaporizer, or incorrect full load mixture adjustment. Restricted propane filters cause frosting of the filter.

HESITATION ON ACCELERATION

1. Check ignition timing and distributor advance.
2. Test CO levels for lean mixtures. Air leaks into the mixer or intake manifold could cause lean air-fuel mixtures.
3. On Garretson systems, excessive venturi plate opening or excessive secondary valve spring tension will result in lean mixtures and a hesitation on acceleration.

MISFIRING ON ACCELERATION

1. Check cylinder compression.
2. Check spark plug heat range.
3. Be sure coil polarity is correct.
4. Test spark plugs, plug wires, and distributor cap.
5. Check spark plug wire routing as outlined in Chapter 4.

LOW FUEL ECONOMY

1. Prove fuel economy; be sure customer record is accurate.
2. Check cylinder compression.
3. Check ignition timing and distributor advances.
4. Test CO levels.

DETONATION AND PINGING

1. Check ignition timing and distributor advances.
2. Check distributor cap and spark plug wires for proper routing.
3. An enclosed air cleaner with a cold air intake will reduce detonation, especially on heavy duty applications.

────────────── Questions ──────────────

1. Secondary ignition voltage requirements are higher under hard acceleration than idle speed voltage requirements. T F
2. List three defects that could lower the maximum available secondary voltage.
    a. _____
    b. _____
    c. _____
3. A propane-fueled engine has higher secondary voltage requirements than a gasoline-fueled engine. T F
4. The spark plug gaps should be _____ when an engine is converted from gasoline to propane.
5. Normal spark plug carbon conditions are _____ in color on a propane-fueled engine.
6. When spark plug operating temperatures are lowered, the voltage requirements of the spark plug will decrease. T F
7. The centrifugal advance mechanism rotates the _____ _____ in relation to the direction that the distributor shaft is rotated.
8. Many vacuum advance diaphragms have an adjusting nut inside the vacuum outlet. T F
9. A propane-fueled engine requires more spark advance at low rpm and less spark advance at high rpm than a gasoline-fueled engine. T F
10. Excessively late ignition spark advance could result in damage to the exhaust valves. T F
11. The ideal propane air-fuel ratio is _____.
12. Carbon monoxide levels should decrease when the engine is accelerated from idle to normal cruise conditions. T F
13. Lean propane air-fuel ratios can result in hard starting. T F
14. A CO level of 1 percent at wide open throttle would result in a loss of power. T F

CHAPTER 6

Propane Conversions and Automotive Computer Systems

A Computer Command Control (3C) System

PURPOSE Many newer vehicles are equipped with dual phase catalytic converters. A platinum or palladium converter lowers HC and CO emission levels. Nitrous oxide (NOx) emissions are reduced by a rodium converter. To control all three pollutants, a dual phase converter requires an exact air-fuel ratio of 14.6:1. Lean air-fuel ratios provide low HC and CO levels, and converter efficiency in correcting these emission levels is high. High cylinder temperatures and increased nitrous oxide emissions result from lean air-fuel ratios, and converter efficiency in correcting NOx emissions is low, as illustrated in Figure 6-1.

Rich air-fuel ratios provide low levels of NOx and high converter efficiency in correcting NOx emissions. High emissions of HC and CO result from rich air-fuel mixtures, and converter efficiency in correcting these emission levels is extremely low. If the air-fuel ratio is kept in a narrow range near 14.6:1, both converters are able to operate at approximately 90 percent efficiency in correcting all three pollutants. Computer systems are necessary to provide rigid control of air-fuel ratios.

COMPUTER INPUT SIGNALS An exhaust gas oxygen (EGO) sensor is located in one of the exhaust manifolds. Rich air-fuel ratios supply fuel to mix

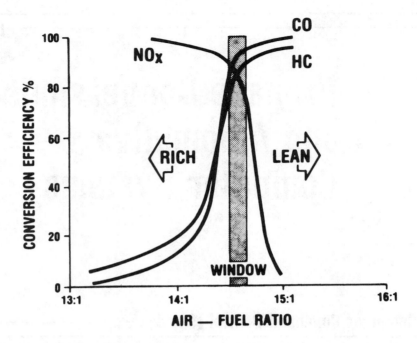

FIGURE 6-1. Catalytic Converter Efficiency. *(Courtesy of Sun Electric Corporation)*

with all the oxygen entering the engine; because excess fuel mixes with all the oxygen entering the engine, oxygen levels in the exhaust remain low. High exhaust oxygen levels result from lean mixtures owing to the lack of fuel entering the cylinders. Exhaust gas is applied to the outside of the EGO-sensing element and atmospheric pressure is supplied to the inside of the sensing element, as shown in Figure 6-2.

Rich mixtures and low exhaust oxygen levels cause the EGO-sensing element to generate higher voltage, as pictured in Figure 6-3. High exhaust oxygen levels and lean mixtures result in low EGO sensor voltage because high oxygen levels are present on both sides of the sensing element. The EGO sensor signal to the computer is used to control the air-fuel mixture in the carburetor.

THROTTLE POSITION SENSOR (TPS) The throttle position sensor is a variable resistor operated by the accelerator pump linkage, as illustrated in Figure 6-4. A reference voltage is supplied to the TPS from the computer. The TPS output varies in relation to throttle opening. At wide open throttle, and on sudden acceleration the TPS signal causes mixture enrichment.

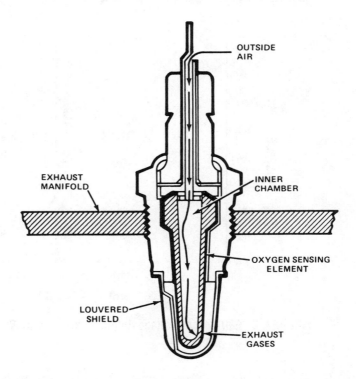

FIGURE 6-2. EGO Sensor. *(Courtesy of General Motors of Canada, Ltd.)*

COOLANT TEMPERATURE SENSOR The resistance of the coolant sensor varies in relation to coolant temperature. The computer operating mode is selected from the coolant sensor signal. Open loop operation occurs during engine warm-up, when a richer mixture is necessary. In the open loop mode the computer maintains the carburetor mixture. Closed loop operation occurs when the coolant sensor reaches a predetermined temperature. The computer controls the air-fuel ratio from the EGO sensor signal in the closed loop mode. The coolant sensor signal may also be used by the computer to assist in ignition management, air injection management, and exhaust gas recirculation management. Figure 6-5 illustrates a coolant temperature sensor.

PRESSURE SENSORS A reference voltage is supplied to the manifold absolute pressure (MAP) sensor from the computer, as illustrated in Figure 6-6. The voltage output signal from the MAP sensor to the computer varies in relation to the manifold vacuum. The computer

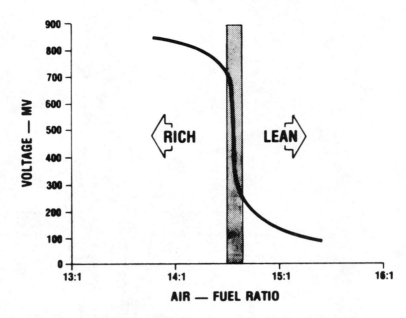

FIGURE 6-3. EGO Sensor Voltage Output. *(Courtesy of Sun Electric Corporation)*

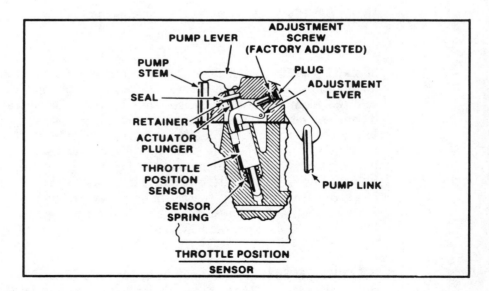

FIGURE 6-4. Throttle Position Sensor. *(Courtesy of General Motors of Canada, Ltd.)*

Propane Conversions and Automotive Computer Systems 119

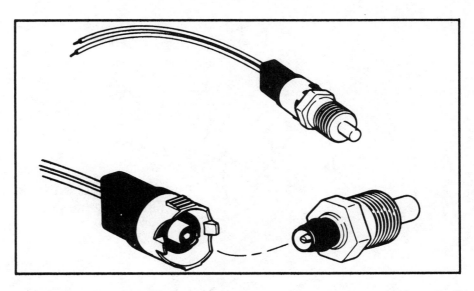

FIGURE 6-5. Coolant Temperature Sensor. *(Courtesy of General Motors of Canada, Ltd.)*

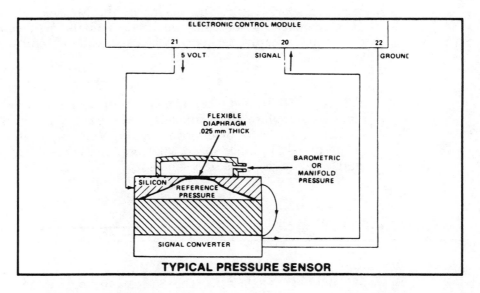

FIGURE 6-6. Manifold Absolute Pressure Sensor. *(Courtesy of General Motors of Canada, Ltd.)*

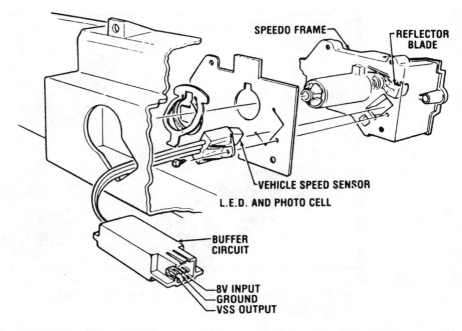

FIGURE 6-7. Vehicle Speed Sensor. *(Courtesy of General Motors of Canada, Ltd.)*

manages spark control by means of the MAP sensor signal. Other management functions may be affected by the MAP sensor signal. A barometric pressure sensor may be used with the MAP sensor on some applications.

VEHICLE SPEED SENSOR (VSS) The speedometer cable rotates a reflector blade past a light-emitting diode and photo cell in the VSS, as shown in Figure 6-7. Reflector blade speed and the VSS output signal are directly proportional to vehicle speed. The computer controls torque converter clutch lockup from the VSS signal.

Computer Output Signals

CARBURETOR MANAGEMENT The 3C carburetor contains a computer-controlled mixture control solenoid. Movement of the solenoid plunger controls the metering rods in the main jets and the air-bleed pin in the idle circuit. The mixture control solenoid plunger is spring loaded in the upward position. When the solenoid winding is activated, the plunger will move downward. If the EGO sensor provides a rich signal, the computer will energize the mixture control solenoid and the metering rods will move downward to provide leaner mixtures. The computer will deactivate the mixture control solenoid winding when the EGO sensor signal indicates a lean mixture. When the solenoid winding is deenergized, the plunger and metering rods can move upward, with the result that the air-fuel ratio will become richer. The computer energizes the mixture control solenoid winding approximately 10 times per second to maintain the air-fuel ratio at 14.6:1. With the upward motion of the mixture control solenoid plunger, the idle air-bleed pin will also move upward; air flow into the idle circuit will then be reduced and idle mixtures will be richer. Most defects in the EGO sensor, mixture control solenoid winding, or computer will result in upward movement of the mixture control solenoid plunger and excessively rich air-fuel mixtures. Figure 6-8 illustrates a 3C carburetor.

IGNITION MANAGEMENT The 3C high-energy ignition system utilizes a seven-terminal module, as pictured in Figure 6-9. When the engine starts, the pickup coil signal goes directly through the module signal converter and bypass circuit. The electronic spark timing (EST) circuit in the computer has no effect on spark advance while the engine is cranking.

The 5-v disable circuit is activated when the engine starts, and the module bypass circuit across the lower bypass contacts is completed. The pickup coil signal now travels through the module signal converter and the crankshaft position lead to the EST circuit in the computer. A modified pickup signal is sent from the computer through the compensated ignition spark lead to the HEI module. The computer will modify the pickup coil signal on the basis of the input signals received, and will provide the exact spark advance required by the engine. When basic timing is being checked, the four-wire distributor connector must be disconnected.

TORQUE CONVERTER CLUTCH (TCC) MANAGEMENT The torque converter clutch eliminates slippage in the torque converter under normal cruise driving conditions. A friction disc is splined to the converter turbine, as shown in Figure 6-10. Converter lockup is achieved when oil enters the converter hub and forces the friction disc against the front of the converter. In the lockup mode the flywheel is connected through the front of the converter to the friction disc, turbine, and transmission input shaft. The converter can be unlocked by directing oil through the hollow

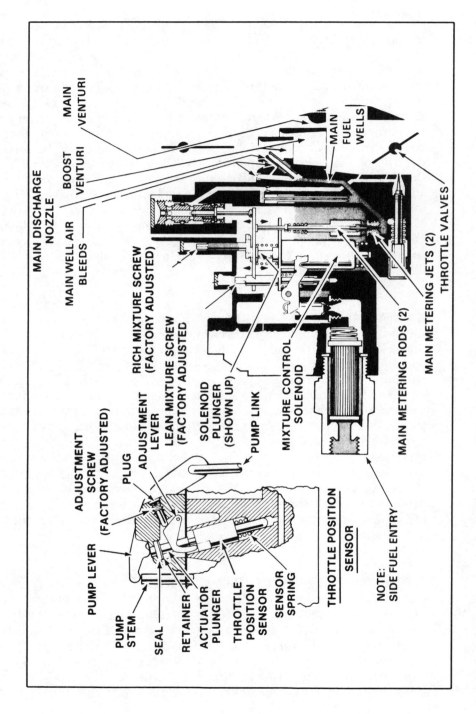

FIGURE 6-8. 3C Carburetor with Mixture Control Solenoid. *(Courtesy of General Motors of Canada, Ltd.)*

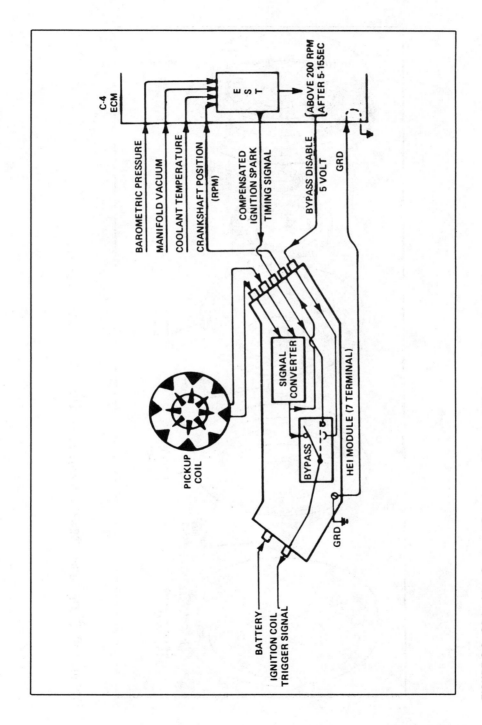

FIGURE 6-9. 3C Ignition System. *(Courtesy of General Motors of Canada, Ltd.)*

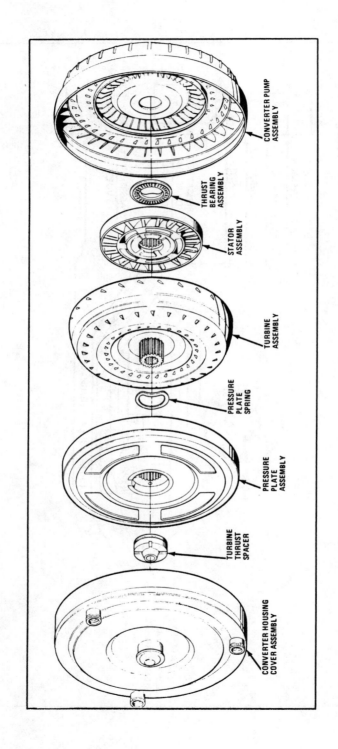

FIGURE 6-10. Lockup Torque Converter. *(Courtesy of General Motors of Canada, Ltd.)*

input shaft and forcing the friction disc away from the front of the converter. Oil flow to the converter is controlled by an apply valve in the transmission. The apply valve is controlled by the TCC solenoid.

When the vehicle reaches a predetermined speed in high gear, the computer grounds the TCC solenoid winding. Energizing of the TCC solenoid causes a hydraulic bleed port in the direct clutch circuit to close and the direct clutch oil pressure to rise. The increase in oil pressure moves the converter apply valve and directs oil in the converter hub to lock up the converter. Other input signals to the computer may affect TCC lockup, as indicated in Figure 6-11.

EARLY FUEL EVAPORATION (EFE) MANAGEMENT In the EFE system, the heat riser valve is operated by a vacuum diaphragm, as shown in Figure 6-12.

A cold coolant sensor signal causes the computer to energize the EFE solenoid and apply a vacuum to the power actuator. When the heat riser valve is closed, the intake manifold becomes heated since exhaust gas is forced through the intake manifold crossover passage. At a specific coolant temperature, the computer will deactivate the EFE solenoid and allow the heat riser valve to open. The EFE vacuum is also supplied to the air injection system, as illustrated in Figure 6-13.

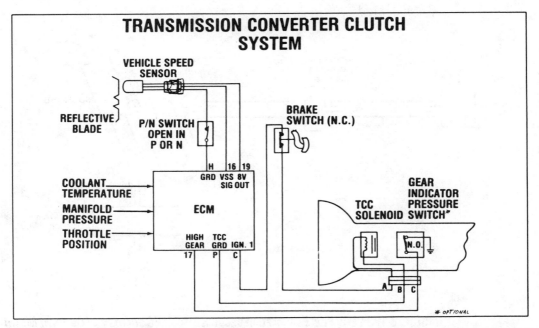

FIGURE 6-11. Torque Converter Clutch Circuit. *(Courtesy of General Motors of Canada, Ltd.)*

Alternate Automotive Fuels

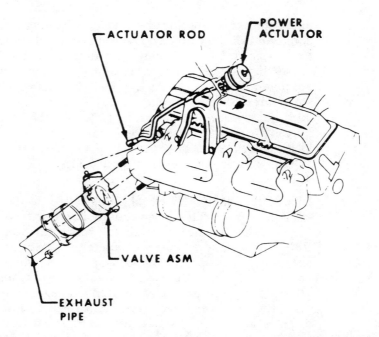

FIGURE 6-12. Early Fuel Evaporation System. *(Courtesy of General Motors of Canada, Ltd.)*

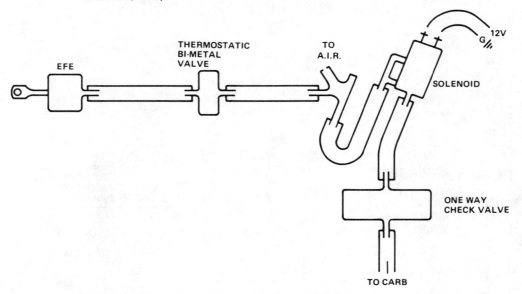

FIGURE 6-13. Early Fuel Evaporation Vacuum Circuit. *(Courtesy of General Motors of Canada, Ltd.)*

AIR INJECTION REACTOR (AIR) MANAGEMENT The AIR system supplies air from the air pump to the exhaust ports or to the catalytic converters, as illustrated in Figure 6-14. Air flow to the exhaust ports lowers emission levels during warm-up and reduces EGO sensor and converter warm-up time. Oxygen is necessary for converter operation once the converters have reached operating temperature.

A divert valve and a switching valve are used to control air flow from the AIR pump. During engine warm-up, the computer deenergizes the divert valve solenoid winding and allows AIR pump pressure to move the divert valve diaphragm and spool valve upward. As indicated in Figure 6-15, air from the pump is diverted to the air cleaner when the spool valve is in the upward position.

Computer activation of the divert valve solenoid winding shuts off AIR pump pressure to the diaphragm and allows the diaphragm and spool valve to move downward. With the divert valve spool in the downward position, air is directed from the AIR pump to the air-switching valve. A high manifold vacuum on deceleration can lift the divert valve diaphragm and cause air to exhaust to the air cleaner.

Cold engine coolant causes the EFE vacuum to be applied to the lower air-switching valve vacuum port. When the vacuum is applied to the lower vacuum port, air will stop flowing from the AIR pump to the air-switching valve diaphragm, and the diaphragm and spool valve will

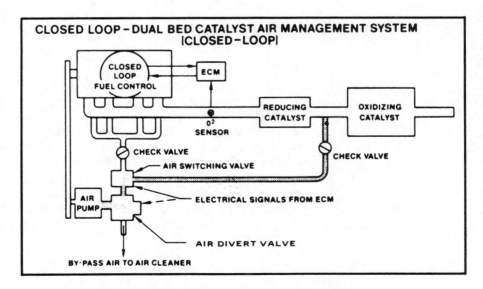

FIGURE 6-14. Air Injection Reactor System. *(Courtesy of General Motors of Canada, Ltd.)*

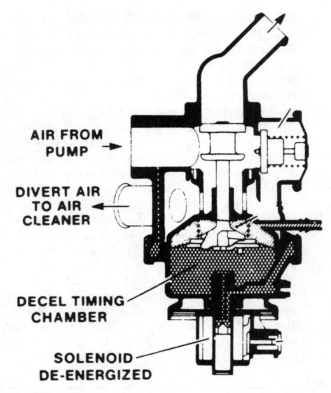

FIGURE 6-15. AIR System Divert Valve. *(Courtesy of General Motors of Canada, Ltd.)*

remain in the downward position. Air from the AIR pump is directed to the exhaust ports with the spool valve in the downward position. Since the vacuum to the lower air-switching valve port is shut off when the coolant is warm, the AIR pump pressure forces the diaphragm and spool valve upward. When the spool valve is in the upward position, the flow of air through the air-switching valve will be directed to the converters, as indicated in Figure 6-16.

If the manifold vacuum is applied to the upper air-switching valve port, the diaphragm and spool valve will be lifted on deceleration with a cold engine. Air will be directed momentarily to the converters to prevent manifold backfiring.

EXHAUST GAS RECIRCULATION (EGR) MANAGEMENT The EGR valve directs exhaust from the exhaust system to the intake manifold and lowers nitrous oxide emissions. The vacuum to the EGR valve is controlled by a computer-operated solenoid. The computer shuts off the EGR vacuum during cold coolant conditions by energizing the solenoid. Warm engine coolant conditions signal the computer to deactivate the solenoid so that the vacuum can be applied to the EGR valve. The vacuum port connected to the EGR system is always above the throttles. The vacuum should never be applied to the EGR valve until the throttles are opened. An EGR valve that is open at idle speed will cause rough idling. Figure 6-17 illustrates the EGR valve and the vacuum control circuit.

IDLE SPEED MANAGEMENT The computer and idle speed control motor determine the idle rpm under all conditions. Cold coolant conditions signal the computer to increase idle speed. Activation of the air-conditioning clutch will signal the computer to maintain idle speed at the specified rpm. When the transmission selector is moved from drive to park, the computer will reduce idle speed to specifications. The idle speed control motor circuit is shown in Figure 6-18.

SELF-DIAGNOSTICS A check-engine light on the instrument panel is energized by the computer if certain defects occur in the system. Service

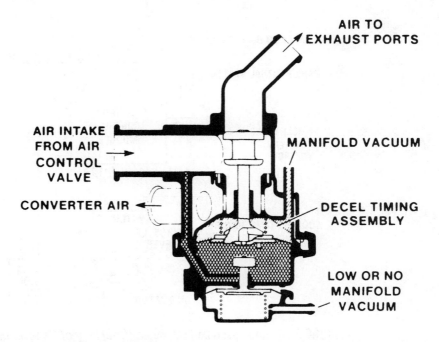

FIGURE 6-16. Air-Switching Valve. *(Courtesy of General Motors of Canada, Ltd.)*

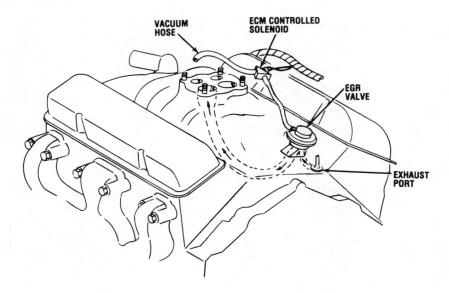

FIGURE 6-17. Exhaust Gas Recirculation Valve. *(Courtesy of General Motors of Canada, Ltd.)*

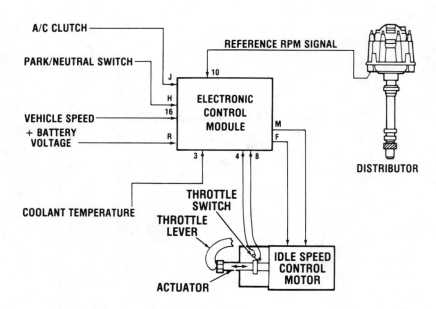

FIGURE 6-18. Idle Speed Control Motor. *(Courtesy of General Motors of Canada, Ltd.)*

Propane Conversions and Automotive Computer Systems 131

personnel can ground a computer diagnostic lead wire in an under-dash connector, as indicated in Figure 6-19. Various diagnostic connectors are used, depending on the year and make of vehicle. The check-engine light should be on for a few seconds each time the engine is started. Sometimes the engine has to be operated at part throttle for five minutes before the check engine light will indicate a system defect.

When the diagnostic lead is grounded, the check-engine light will begin flashing defective codes stored in the computer, as indicated in Figure 6-20. One flash followed by a pause and two more flashes in quick succession indicates code twelve. The diagnostic codes are given in numerical order and repeated three times.

Defective codes may be erased from the computer memory by disconnecting the continuous 12-v supply wire to the computer for about 2 minutes. Figure 6-21 indicates the possible defective codes. Code 44 would be caused by an excessively lean EGO signal. Diagnostic codes may vary, depending on the year of vehicle.

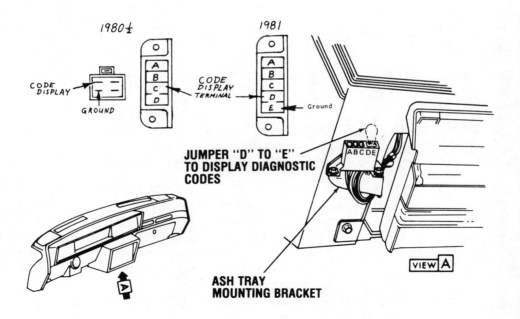

FIGURE 6-19. Self-Diagnostic Connector. *(Courtesy of General Motors of Canada, Ltd.)*

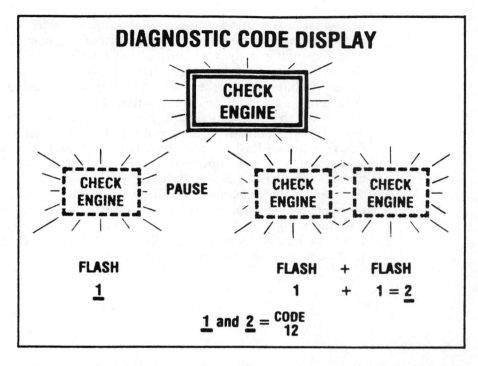

FIGURE 6-20. Check-Engine Light Defective Codes. *(Courtesy of General Motors of Canada, Ltd.)*

Precautions For Propane Conversions On Computer Systems

1. Do not disconnect any of the system components. For example, if the air pump is disconnected, the EGO sensor signal might be affected under certain operating conditions. When some components are disconnected, the computer will turn on the check-engine light. The mixture control solenoid will remain in the full rich position if the EGO sensor is inoperative.
2. Testing the EGO sensor with a voltmeter or shorting the terminals could destroy the sensor.
3. The carburetor bowl and air horn assembly must not be discarded when installing a straight propane conversion. A mixer must be

TYPICAL SELF-DIAGNOSTIC FLASHING CODES INDEX

DIAGNOSTIC CODE	MALFUNCTION DESCRIPTION
12	NO TACHOMETER OR DISTRIBUTOR PULSES TO ECM
13	OXYGEN SENSOR CIRCUIT
14	SHORTED COOLANT SENSOR CIRCUIT (HIGH COOLANT TEMP.)
15	OPEN COOLANT SENSOR CIRCUIT (LOW COOLANT TEMP.)
21	THROTTLE POSITION SENSOR CIRCUIT FAILED HIGH, OR WOT
21	THROTTLE POSITION SENSOR ADJUSTMENT, OR STICKING
23	OPEN OR GROUNDED CARBURETOR SOLENOID
24	VEHICLE SPEED SENSOR CIRCUIT
32	BAROMETRIC SENSOR CIRCUIT LOW, OPEN OR GROUNDED
34	MANIFOLD PRESSURE SENSOR CIRCUIT LOW, HIGH, OPEN OR GROUNDED
34	MANIFOLD PRESSURE SENSOR HOSE OFF
35	THROTTLE SWITCH SHORT (ISC)

DIAGNOSTIC CODE	MALFUNCTION DESCRIPTION
44	LEAN OXYGEN SENSOR
45	RICH OXYGEN SENSOR
51	PROM — CALIBRATION UNIT
52	ECM FAILURE
53	ECM FAILURE
54	CARB. SOLENOID CIRCUIT
55	ECM FAILURE

FIGURE 6-21. 3C Diagnostic Codes. *(Courtesy of General Motors of Canada, Ltd.)*

INSTALLATION

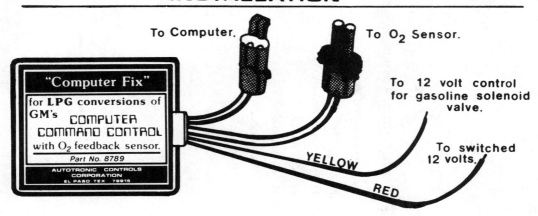

FIGURE 6-22. Computer Fix Control. *(Courtesy of Autotronic Controls Corporation)*

adapted to the top of the gasoline carburetor, and the gasoline fuel pump discarded on a straight propane conversion.

4. The check-engine light may come on in the propane mode because of reduced oxygen levels in the exhaust, especially under cruise conditions. The computer fix control may be installed to bypass the EGO sensor in the propane mode and to eliminate the check-engine light problem. Normal EGO sensor operation is allowed by the computer fix control in the gasoline mode. A computer fix control for systems with two EGO sensor terminals is illustrated in Figure 6-22. Controls are available for systems with single EGO sensor terminals.

Questions

1. When a dual phase catalytic converter is used, the gasoline air-fuel ratio must be maintained at _____.
2. The EGO sensor generates a higher voltage when the air-fuel ratio is lean.   T   F
3. The computer selects open loop or closed loop operation from the _____ sensor signal.
4. The vehicle speed sensor is operated by the _____.

5. When the computer receives a rich EGO sensor signal, it energizes the mixture control solenoid winding.   T   F
6. The computer and the mixture control solenoid will control the air-fuel mixture at all speeds.   T   F
7. Ignition spark advance is controlled by the computer while the engine is cranking.   T   F
8. The vehicle speed sensor signal is used by the computer in _____ management.
9. The EGR valve should be open at idle with a warm engine.   T   F
10. The EGO sensor should be tested with an ordinary voltmeter.   T   F
11. When a C3-equipped vehicle is converted to propane, the AIR pump should be disconnected.   T   F

CHAPTER 7

Natural Gas—The Answer To Inexpensive Fuel

Natural Gas Facts

AVAILABILITY In many nations of the world natural gas is becoming increasingly available. The use of liquid natural gas (LNG) tankers to transport natural gas has increased drastically since 1970. A typical LNG tanker carrying 750,000 bbl (125,000 m^3) of LNG can fuel 10,000–20,000 vehicles for a year. Thailand's pipeline carrying natural gas 264 mi (422 km) from the Gulf of Thailand to Bangkok is the longest submarine pipeline in the world. Bangkok's streets are choked with smog-creating vehicles that burn gasoline refined from imported crude oil. If these vehicles were powered with natural gas, the environment and the economy in Thailand could improve significantly.

A natural gas pipeline spanning the Mediterranean supplies 1.2 billion ft^3 (34.2 million m^3) per day from Algeria to Italy, as illustrated in Figure 7-1.

The Algerian pipeline supplies natural gas for Italy's 250,000 vehicles powered with compressed natural gas (CNG) and meets commercial natural gas requirements.

The giant Maui and Kapuni gas fields off the coast of New Zealand contain 6 trillion ft^3 (170 billion m^3) of proven reserves. Some of the locations of automotive CNG refueling stations that have been constructed in New Zealand are shown in Figure 7-2. About 25,000 vehicles have been converted from gasoline to CNG in New Zealand.

Many other nations have natural gas reserves and distribution systems, as indicated in Table 7-1. Sources of synthetic natural gas such as coal gasification, kelp, and biomass are used by some nations. The increasing availability of LNG on a worldwide basis will probably encourage the use of natural gas.

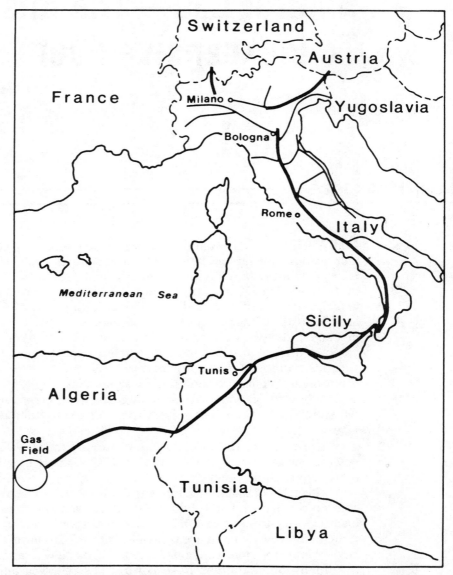

FIGURE 7-1. Mediterranean Natural Gas Pipeline. *(Courtesy of CNG Fuel Systems, Ltd.)*

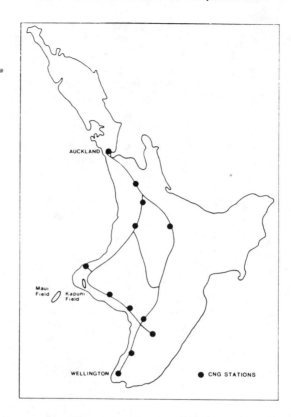

FIGURE 7-2. New Zealand CNG Refueling Station Locations. *(Courtesy of CNG Fuel Systems, Ltd.)*

AVAILABILITY IN THE UNITED STATES The 1-million-mile U.S. natural gas pipeline system carries gas from oil and gas fields in the Southwest to the Northeast and Midwest. The annual capacity of the installed natural gas pipeline system is estimated at 23 trillion ft^3 (650 billion m^3). The present annual demand of 20 trillion ft^3 (570 billion m^3) leaves a reserve pipeline capacity of 13 percent. Approximately 14 trillion ft^3 (400 billion m^3) of natural gas would be required annually to fuel the entire fleet of highway vehicles. If this entire fleet was converted to CNG, the capacity of the pipeline to supply the fuel would be far exceeded. The 13 percent reserve pipeline capacity could supply about 20 percent of the fleet with CNG. As mentioned in Chapter 1, the United States has huge natural gas reserves in geopressured zones and large reserves of coal that could be converted to synthetic natural gas. These unconventional and synthetic sources of natural gas will have to be utilized if a significant percentage of the vehicle fleet is converted to CNG or LNG.

TABLE 7-1. Natural Gas Reserves

Country	Proven Reserves (TCF)	Reserves-To-Production Ratio (R/P)
Algeria	131	348
Argentina	22	50
Australia	30	105
Bahrain	9	107
Bangladesh	8	*
Bolivia	6	49
Canada	87	21
China	30	9
Colombia	8	41
Ecuador	4	25
Egypt	3	33
France	6	23
Germany, West	6	9
India	12	64
Indonesia	24	24
Iran	485	548
Iraq	28	570
Italy	8	18
Kuwait	31	152
Libya	24	149
Malaysia/Brunei	26	78
Mexico	65	33
Netherlands	62	19
New Zealand	6	152
Nigeria	41	115
Norway	43	32
Pakistan	15	65
Qatar	60	1476
Saudi Arabia	110	450
Thailand	8	v. large*
Trinidad/Tobago	12	49
Tunisia	6	*
United Arab Emirat	22	340
United Kingdom	25	21
United States	191	10
USSR	920	63
Venezuela	42	101

SOURCE: Courtesy CNG Fuel Systems Ltd.
*Data not available.

AVAILABILITY IN CANADA Thirty percent of Canada's crude oil requirements are imported, while natural gas reserves have been increasing each year. In recent years Canadian reserves have increased as much as 5 trillion ft^3 (140 billion m^3) annually. National consumption has been about 1.6 trillion ft^3 (45.7 billion m^3) annually, and exports have averaged 1 trillion ft^3 (28.5 billion m^3) per year. If 500,000 vehicles were converted to CNG, they would consume 0.1 trillion ft^3 (2.8 billion m^3) each year. Approximately 200,000 bbl (33,000 m^3) of crude oil per day are required to provide gasoline for the same number of vehicles. The existing and proposed natural gas pipelines in Canada are outlined in Figure 7-3.

COST FACTORS Compressed natural gas will cost 1/3–1/2 the price of gasoline. The following table compares gasoline prices and natural gas prices in different locations of the United States. The figures are based on 1 gal of gasoline being equivalent to 100 ft^3 of natural gas.

	Gasoline (per gal)	CNG (per gal equivalent)
Baltimore	1.25	0.38
Boston	1.30	0.48
Chicago	1.30	0.35
Kansas City	1.20	0.25
San Francisco	1.25	0.30

Vehicle fleets with private filling facilities will probably have less expensive natural gas than vehicles refueling at public filling facilities. A vehicle that is driven 50,000 mi (80,000 km) per year and provides 15 mpg would have a fuel bill of $4,166 per year if gasoline was priced at $1.25 per gallon. CNG motor fuel would provide an annual saving in excess of $2,000. Conversion costs for CNG would be about $1,400–2,200, depending on the number of cylinders installed and the type of conversion equipment used. Natural gas costs and conversion costs vary considerably throughout North America. Pay-back periods for CNG conversions will have to be worked out for each state or province on the basis of local fuel and conversion costs. Additional cost savings would be obtained from reduced oil and filter changes, longer tune-up intervals, and increased engine life.

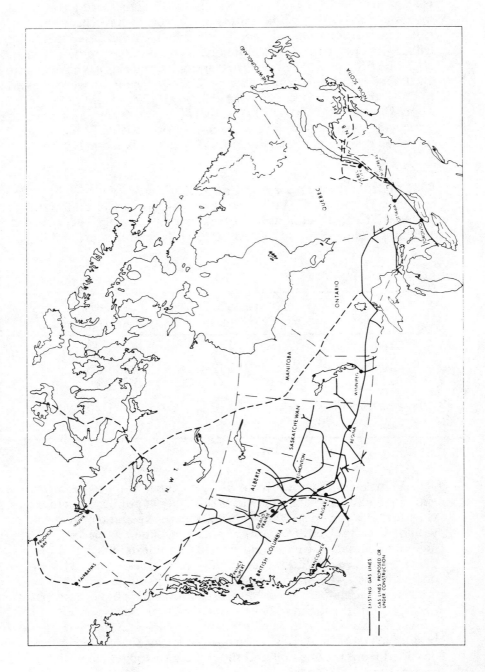

FIGURE 7-3. Natural Gas Pipelines in Canada. *(Courtesy of CNG Fuel Systems, Ltd.)*

COMPOSITION AND TUNING REQUIREMENTS Natural gas is composed mainly of methane, CH_4. Hydrocarbons having a higher carbon to hydrogen ratio, such as ethane, propane, and butane, may also be present in natural gas. These hydrocarbons can make up as much as 20 percent of natural gas. The percentage of higher hydrocarbons varies in each gas field. A small amount of processing is generally required before the natural gas is distributed by pipeline. Natural gas processing includes removal of impurities, higher hydrocarbons, and water vapor.

Natural gas contains 1,000-1,100 BTUs (1,055-1,160 kj) per ft^3. CNG contains 32,154 BTUs (33,332 kj) per U.S. gal at 3,000 psi (210 atm). A power loss of 10-15 percent has been experienced in some gasoline-to-CNG conversions. Natural gas with an octane rating of 130 would allow the use of much higher compression ratios. If engines were optimized for CNG, the power loss would be minimized or eliminated.

IGNITION REQUIREMENTS Timing requirements for CNG fuel differ from gasoline timing requirements. For CNG fuel, basic ignition timing and distributor advance should be increased at low rpm. Because of the reduced range of the CNG cylinders, most CNG conversions will be dual fuel systems. Permanent alteration of distributor advances and ignition timing may cause detonation problems in the gasoline mode. The dual curve ignition control for CNG and gasoline automatically provides the correct CNG spark timing. LPG and gasoline dual curve ignition control units were shown in Chapter 5.

Dynamometer testing can be used to determine the optimum basic timing setting for CNG when the dual curve ignition control unit is not installed. Dynamometer test results of optimum timing settings to minimize engine power loss are illustrated in Table 7-2. In comparison with gasoline, CNG fuel needs additional spark advance at idle and lower speeds, but the same total spark advance at high speeds. If the basic timing is set ahead 5 degrees for CNG fuel, the maximum centrifugal advance should be reduced 5 degrees to maintain the same total spark advance. (Methods of recalibrating distributor advances were discussed in Chapter 5.)

Some types of electronic ignition systems, such as capacitor discharge ignition, give a CNG fueled engine distinct advantages compared to a gasoline fueled engine with point-type ignition. As Figure 7-4 shows, engine torque improves considerably at high rpm on a CNG fueled engine with capacitor discharge ignition compared to a gasoline fueled engine with point-type ignition. (Ignition requirements for propane fuel were outlined in Chapter 5.) CNG and propane ignition requirements are similar. Spark plug gaps for CNG fuel should be the

TABLE 7-2. Dynamometer Test Results for Optimum Timing: Average Power Losses With CNG Operation to Gasoline (Expressed as a percentage of power compared with gasoline)

VEHICLE; TIME SETTINGS	1,000 rpm	1,500 rpm	2,000 rpm	2,500 rpm	3,000 rpm	3,500 rpm	4,000 rpm	4,500 rpm
1. Ford Escort 1.3 Van Timing Settings								
0° BTC	27.25	32	28	31.5	35.8	34.5	29.7	28.7
4° BTC	14.8	13.9	16.7	23.3	28	29.8	24.3	27.7
*8° BTC	10.5	11.8	10	12.7	10.3	8.8	10	8
12° BTC	17.0	19.3	21.8	29.8	18.7	20.3	25	30
2. Marina 400 kg Van Timing Settings								
0° BTC	48.3	43.8	45.8	33.3	22.3	1.0	+6	+12
5° BTC	27.3	24	21.8	15	8	2.0	+10.5	+3
*10° BTC	21	10.3	8	10	2.2	+7.6	+26.3	+15
15° BTC	15	16	15.5	16.5	17.5	18.3	22.5	12

3. **Marina Sedan**
   Timing Settings

5° BTC	19.8	23.5	27	27.3	22	17.5	+12.4	+ 4.3
*10° BTC	11.5	22.5	25.3	28.3	29.8	17	15	18.3
15° BTC	7.3	15	18.8	22.3	29	14.3	12.7	11

4. **Toyota Corolla Sedan**
   Timing Settings

0° BTC	21.5	22.3	24	41.8	27.3	24.3	24.7	34.3
5° BTC	2.0	3.3	32.5	9.5	17.7	31.3	44.3	33.0
*10° BTC	1.0	8.3	11.8	22.8	10.3	10.5	11.0	+12.0
15° BTC	6.0	18.0	28.0	32.3	36.0	20.8	22.3	38.5

5. **Ford Transit Van**
   Timing Settings

8° BTC	23.75	18.5	20.67	19	12.33	+ 7.75	-	-
*12° BTC	16.25	21.38	11	7.25	8.75	3.5	-	-

* Optimum timing for engine power with C.N.G.

SOURCE: Courtesy New Zealand Energy Research and Development Committee.

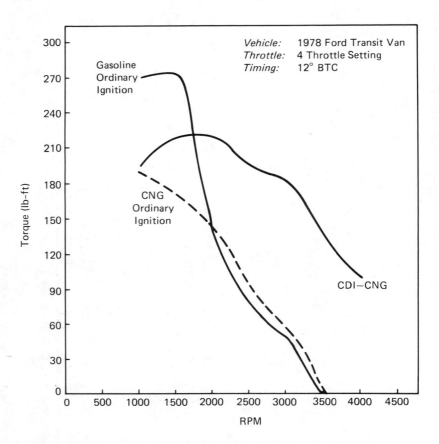

FIGURE 7-4. Engine Torque with CD and Point Ignition. *(Courtesy of New Zealand Energy Research and Development Committee)*

same as for propane fuel. Secondary coil polarity and spark plug wire routing are also important in CNG-fueled engines.

CARBON MONOXIDE LEVELS The ideal CNG air-fuel ratio is 16.5:1. Carbon monoxide levels will be about 1 percent under the ideal CNG air-fuel ratio, as illustrated in Table 7-3. When an infrared tester is used, CO levels at idle should be about 1 percent. Cruise mixtures should provide slightly lower CO levels in relation to the CO reading at idle. Full throttle CO levels should be about 3 percent. In some areas CO levels have to be set according to emission laws. (Infrared testing was discussed in Chapter 5.)

TABLE 7-3. Natural Gas Air-Fuel Ratios and CO Levels

Natural Gas	–	18.0	*16.2*	14.8	14.3	13.8	–
Butane Propane	18.0	16.3	*15.4*	14.2	13.6	12.8	12.0
Gasoline	16.0	15.0	14.0	*13.0*	12.0	11.0	10.0
Carbon Monoxide %		0	1	2	3	4	5 **6**	–

EMISSION LEVELS The main pollutants in the exhaust of a gasoline engine are carbon monoxide, unburned hydrocarbons, and oxides of nitrogen. All three pollutants are reduced significantly when CNG is used, as illustrated in Figure 7-5.

U.S. Experience With CNG

FLEET EXPERIENCE More than fifty companies in the United States operate part or all their fleet on CNG. Privately owned vehicles have not been converted to CNG, mainly because of the lack of refueling facilities. The Southern California Gas Company has approximately 2,400 vehicles, of which about 50 percent have been converted to CNG. Vehicles converted to CNG include forklifts, passenger cars, small trucks, and trucks up to 12 tons gvw. The average yearly travel per vehicle is 9,600 mi (14,400 km). Passenger car installations include two 322-ft^3 (9.2-m^3) cylinders in the trunk. Four to six cylinders are installed on larger trucks. The CNG experience of the Southern California Gas Company could be summarized as follows:

1. Engine power loss with correct CNG engine tuning amounted to 15 percent.
2. Fuel costs, engine wear, and emission levels were drastically reduced.
3. Valve wear on CNG vehicles did not progress at a rate that significantly increased the cost of valve repairs.
4. Oil change intervals were doubled with CNG.
5. CNG vehicles were involved in a number of traffic accidents, but CNG componentry did not contribute to any accidents, and no CNG component was considered dangerous as a result of an accident.

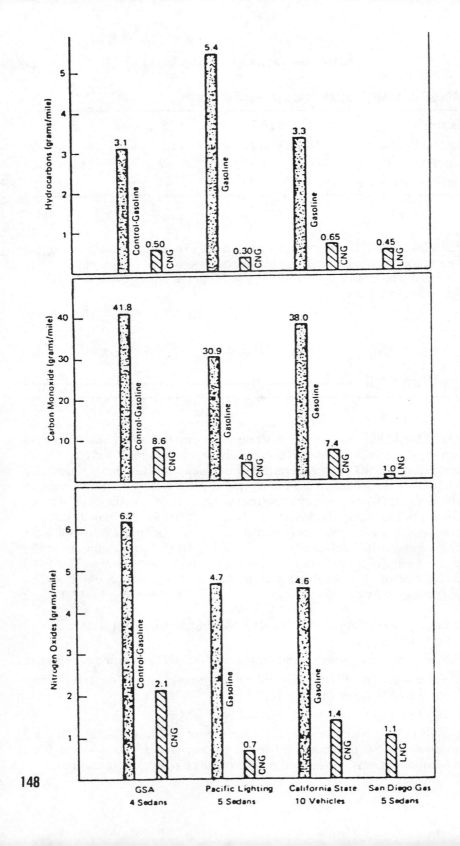

FIGURE 7-5. Emission Levels of Gasoline and CNG. *(Courtesy of New Zealand Energy Research and Development Committee)*

The Arizona Public Service Company discovered the following advantages of using CNG in their fleet:

1. An average reduction of 90 percent in CO and HC emissions were provided with CNG.
2. Nitrous oxide emissions were lowered by 70 percent.
3. Maintenance costs were reduced, particularly those for oil and filter changes.
4. CNG operation resulted in a saving of one major tune-up per vehicle per year.

A number of disadvantages were also experienced:

1. Drivers complained of inconvenience because of insufficient filling facilities in the area.
2. Three CNG storage tanks of 322 ft^3 (9 m^3) only provided a driving range of 133-160 mi (200 to 240 km).
3. Engine power loss averaged 15-20 percent.

Italian CNG Experience

GENERALIZATIONS Italy has more CNG-powered vehicles than any other country in the world. Therefore, the Italian CNG experience is important to any other nation that is considering large numbers of gasoline-to-CNG conversions. A network of 240 filling stations services approximately 250,000 CNG-powered vehicles. Italians have had more than thirty years of experience with CNG as a motor fuel. Most CNG-powered vehicles are privately owned. Central and northern Italy have 8,666 mi (13,000 km) of natural gas pipelines. Current consumption of natural gas for transportation purposes is 10.5 billion ft^3 (300 million m^3) per year. Natural gas consumed by the automative fleet is 2 percent of the total yearly consumption. Forty percent of the CNG filling stations are travasi facilities located beyond the natural gas mains. Travasi stations use large trailers loaded with CNG cylinders. The cylinders are filled at a central compressor station and hauled to the travasi location, where they are used to refuel vehicles. Travasi filling stations were developed in the 1960s to provide an improved CNG filling station network. Compressor stations are replacing travasi stations to some extent.

The use of CNG as an automotive fuel developed because of high gasoline prices and gasoline shortages. Italian experience has demonstrated that gasoline must be about 2½ times the price of CNG before it is economically feasible to convert vehicles to CNG. The use of CNG as a motor fuel has never been actively promoted, but initial development of the refueling network was undertaken by SNAM, the government-owned gas company. Italy imports about 60 percent of its natural gas requirements, and industry has top priority for natural gas. Therefore, the use of CNG for automotive fuel has not been encouraged. The natural gas pipeline from Algeria to Italy will improve natural gas supplies, and the use of CNG motor fuel should increase as a result.

APPROVAL OF CNG CONVERSION EQUIPMENT Italy's Ministry of Transport is empowered to approve all components for CNG conversions. New equipment manufacturers must submit their components to one of the seven ministry testing centers for evaluation purposes. New components or previously approved modified components are inspected and tested on a vehicle for 3,300 mi (5,000 km). After completion of the road test, the components are bench-tested and examined. If components meet all operational tests, inspections, and safety requirements, they are given an approval number that is stamped on these parts and on all subsequent production units. Manufacturers advertising information must quote the approval number.

APPROVAL OF CNG CYLINDERS The Ministry of Transport is responsible for mobile high-pressure equipment. All fixed high-pressure installations such as CNG refueling stations are controlled by the Ministry of Internal Affairs. Two boards have the responsibility for regulations, quality assurance, manufacturing, and testing. All mobile cylinders are the responsibility of the General Inspectorate of Civil Motor Transport (IGMC). Fixed cylinders are the responsibility of the National Association for Control of Combustion (ANCC). The hydrostatic cylinder test pressure is 4,350 psi (300 atm), and the normal filling pressure is 3,200 psi (220 atm). CNG cylinders have an actual burst pressure of at least 8,500 psi (586 atm). All cylinders are stamped by the approving authority.

Italian regulations require retesting of CNG cylinders every five years. The CNG Cylinder Management Fund (CFBM) is responsible for retesting cylinders up to a capacity of 60 l. Testing of CNG cylinders involves ultrasonic crack testing and weighing. The tare weight must not vary more than 6 percent from the original value. Cylinders must be destroyed after thirty years. CFBM owns all CNG cylinders, and the car owner rents the necessary cylinders from CFBM for a nominal fee. If a cylinder fails a five-year test, it is replaced at no cost to the vehicle owner. This system of cylinder rental may be cumbersome, but it has an excellent safety record.

VEHICLE CONVERSION AND TESTING Various automotive workshops in Italy are registered with the Ministry of Transport. These workshops have proved their technical competence in CNG conversions to the ministry. Regulations do not specify who may convert vehicles to CNG but all conversions must be checked and approved by the Ministry of Transport in one of its seven regional testing facilities. A decal is placed on the vehicle after it has received ministry approval. Refueling stations can refuse to fill vehicles without the necessary decal. Transfer of a CNG fuel system to another vehicle requires retesting by the Ministry of Transport. CNG-fueled vehicles must undergo a yearly safety test involving leak testing of all lines and fittings, plus a pressure regulator test.

FILLING STATIONS Privately owned compressor refueling stations must sell 210,000–350,000 ft^3 (6,000 to 10,000 m^3) of CNG per day to be economically feasible. This would involve refueling 200–300 vehicles per day. To be economically profitable, travasi stations must refuel 150–250 vehicles per day and average daily CNG sales must exceed 140,000 ft^3 (4,000 m^3). Compressor stations must have a minimum gas inlet pressure of 145 psi (10 atm). Natural gas compressors operate with much higher efficiency at higher inlet gas pressures. Regulations pertaining to the construction of CNG refueling stations are very strict. Areas required for stations are quite large, 5.6 acres (14 ha). Compressor buildings, gas storage areas, and refueling areas must be separated by 50 ft (15 m). Reinforced concrete must be used in all buildings, including refueling bays. A typical compressor refueling station is illustrated in Figure 7-6.

Two main compressors are used in the filling station. The compressors fill large storage cylinders to 2,900 psi (200 atm). Vehicles are refueled from the storage cylinders. Refueling of vehicles is completed by a smaller boost compressor. Maximum refueling pressure is 3,200 psi (220 atm). Accurate CNG flow meters were not available until recent years. The Italians did not use CNG flow meters in their filling stations because of the cost of meters and the difficulty of obtaining them. The vehicle approval certificate indicates the cylinder capacity. On the basis of this quantity and the pressure readings before and after refueling, the amount of gas dispensed is read from a chart. All filling stations use the same approved chart.

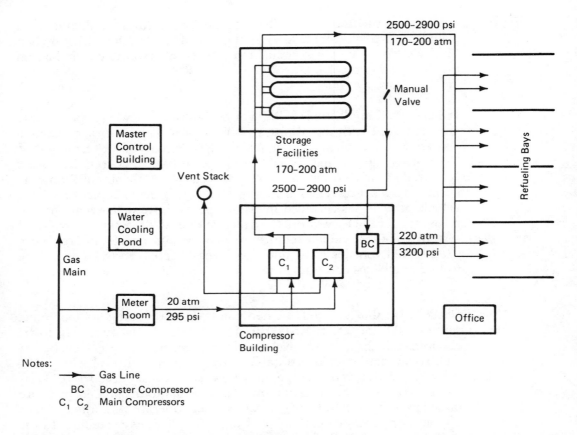

FIGURE 7-6. Italy's Compressor-type CNG Refueling Station. *(Courtesy of New Zealand Liquid Fuels Trust Board)*

CNG Safety Record

GENERAL All Italian organizations concerned agree that no recorded accident has resulted from a faulty CNG system. When CNG-equipped vehicles were involved in accidents, CNG components rarely failed. No recorded deaths or injuries have been attributed to CNG equipment. In some cases, CNG-equipped vehicles have been completely burned out, but no explosive accidents have been recorded. A high level of safety is ensured by cylinder testing at five-year intervals, the checking of all conversions, and the monitoring of manufacturing quality. Strict filling station regulations have also contributed to the excellent safety record. Accident damage involving cylinders must be reported. GFBM, the cylinder management organization, has the responsibility of testing and recertifying cylinders involved in accidents.

In the early years of CNG use, moisture and sulphur impurities in the natural gas caused internal cylinder corrosion that in turn led to some cylinder failures. Particular attention is now given to drying and conditioning the gas to eliminate these corrosive problems.

Italian authorities are opposed to the use of cylinder relief valves or burst discs in motor vehicle cylinders. In their experience, gas released from a burst disc can feed the flames of an existing fire and thus can create an additional hazard. The possibility of cylinder rupture due to excessive heat or mechanical damage is so remote that the use of burst discs is invalidated.

ANCC SAFETY RECORDS All cylinder failures have occurred in very old cylinders in which internal microcracks had developed. Such failures averaged two to three per year when CNG was first used. Compulsory ultrasonic crack testing has reduced the number of cylinder failures. The elimination of moisture and sulphur impurities from the natural gas has also reduced the incidence of cylinder failure. Cylinder failures have never occurred in vehicle collisions. In the case of cylinder failures during refueling, no injuries occurred because of the reinforced concrete refueling bays. Most cylinder failures occur on cylinder trucks where the cylinders undergo extra stress owing to frequent refilling. ANCC do not consider a cylinder failure rate of two or three per year to be excessive—such a failure rate represents only 0.0003 percent of the total number of cylinders in Italy.

SNAM SAFETY RECORDS Since 1970 one accident has occurred involving a vehicle cylinder, and two truck cylinder failures have been recorded. No recorded cylinder failure has resulted from vehicle collisions.

CONCLUSIONS In view of the large number of CNG vehicles (250,000) that have been operating in Italy since 1950, the Italian safety record is excellent. CNG accidents have never resulted in death or injuries. When CNG vehicles were involved in accidents, CNG components did not fail. Three cylinder failures, out of a total cylinder population of 800,000, were recorded from 1970 to 1982. A testimony to the safe record of CNG operation is the current move by regulatory organizations such as SNAM,

ANCC, and the Ministry of Transport to seek a relaxation of some CNG regulations.

―――――――――――― **Questions** ――――――――――――

1. The United States has _____ miles of natural gas pipelines.
2. Reserve capacity in the U.S. natural gas pipeline system is _____ percent.
3. The nation with the most CNG-powered vehicles is _____.
4. A power loss of _____ to _____ percent is encountered on gasoline engines converted to CNG.
5. CNG-powered vehicles require additional spark advance at _____ rpm compared with an engine operating on gasoline.
6. The ideal natural gas air-fuel ratio is 17.5:1.   T   F
7. Levels of nitrous oxide emissions are increased with CNG fuel.   T   F
8. Italian regulations demand that CNG cylinders be tested every _____ years.

CHAPTER 8

Compressed Natural Gas Conversion Equipment

CNG Components

FUEL CYLINDERS CNG fuel tanks are available in various sizes. Some of the tank sizes and equivalent gasoline capacities are listed in Table 8-1. Lighter tanks constructed from high-strength steel are now available. Aluminum tanks wrapped with fiberglass have been introduced recently. A thick wall tank (12 3/4" × 3'9") would weigh about 125 pounds. Aluminum or high-strength steel cylinders weigh 30–50 percent less than the older, thick wall cylinders.

LINES AND FILL FITTING Steel lines are used to connect the tanks to the fill fitting and pressure regulator. The burst pressure rating of the steel lines is 15,000 psi (1,020 atm). Burst discs are located in the tanks and fill fitting. If excessive pressure builds up in the cylinders or fuel lines because of extremely high temperatures, the burst discs will blow rather than rupture the cylinders or fuel lines. The quick-connect fill fitting can be removed and the filler hose inserted in its place. A one-way check valve allows natural gas to flow from the fill hose to the tanks but prevents gas flow from the tanks out of the fill fitting. Figure 8-1 shows an enlarged view of the end of the fill hose that is inserted into the fill fitting.

PRESSURE REGULATORS Tank pressure is lowered to 55 psi (385 kp) by the first-stage regulator. When the first-stage regulator pressure reaches 55

TABLE 8-1. CNG Fuel Cylinder Specifications

Size	Design Pressure	Cubic Feet Water Volume	SCF Natural Gas	Gasoline Equivalent Gallons
12¾" OD × .395 × 3'9"	3000	2.44	634	6.34
12¾" OD × .395 × 4'9"	3000	3.20	831	8.31
12¾" OD × .395 × 5'9"	3000	3.96	1028	10.28
14" OD × .434 × 3'9"	3000	2.90	753	7.53
14" OD × .434 × 4'9"	3000	3.81	989	9.89
14" OD × .434 × 5'9"	3000	4.73	1228	12.28
16" OD × .488 × 3'9"	3000	3.84	997	9.97
16" OD × .488 × 4'9"	3000	5.05	1311	13.11
16" OD × .488 × 5'9"	3000	6.25	1622	16.22

psi (385 kp), the inlet valve will be closed by the upward movement of the diaphragm, as indicated in Figure 8-2. When gas is used out of the first-stage regulator, the pressure will decrease and the inlet valve will be able to reopen. Tank pressure is always present in the first-stage regulator. When the fuel solenoid is energized, fuel vapor will flow through the fuel solenoid to the second-stage regulator. Most CNG conversions are dual fuel systems using a rotary dash switch similar to the LPG gasoline switches described in Chapter 4. The low-pressure regulator valve opens when the mixer vacuum is applied to the regulator diaphragm. The mixer vacuum and low-pressure regulator valve opening increase in relation to engine rpm.

MIXER The mixer illustrated in Figure 8-2 is a tapered gas valve mixer very similar to the tapered gas valve propane mixers described in Chapter 3. The mixer vacuum lifts the tapered gas valve upward in relation to engine speed. An idle mixture adjustment and a full load adjustment are provided on the mixer. The mixer is adapted to the top of the gasoline carburetor.

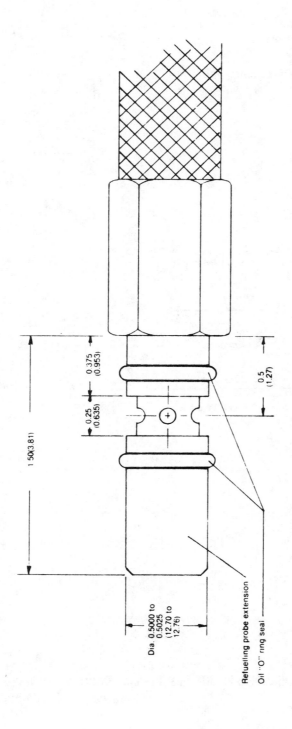

FIGURE 8-1. CNG Filler Hose. *(Courtesy of Canadian Gas Association)*

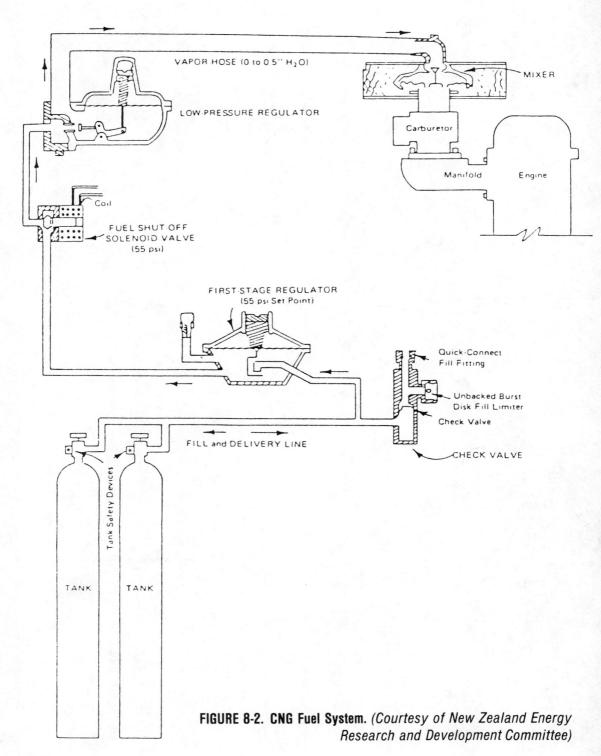

FIGURE 8-2. CNG Fuel System. *(Courtesy of New Zealand Energy Research and Development Committee)*

Tartarini CNG Equipment

PRESSURE REGULATOR The Tartarini pressure reducer is a three-stage unit with a built-in CNG fuelock, prime solenoid, and power valve. Natural gas enters fuel inlet 1 from the fuel cylinders, as illustrated in Figure 8-3.

Fuel passes through filter 2 and past seat 5 into area A. When the pressure in area A reaches 35 psi (245 kp), upward movement of the first-stage diaphragm will close seat 5. CNG flows from area A through seat 7 into area B. The second-stage diaphragm and seat 7 control the pressure to 14.7 psi (100 kp). The mixer vacuum is applied to the third-stage area C. Fuel flows from area B past valve 14 to area C. The mixer vacuum will move diaphragm 23 upward in relation to engine rpm. Upward movement of diaphragm 23 opens valve 14 in relation to engine

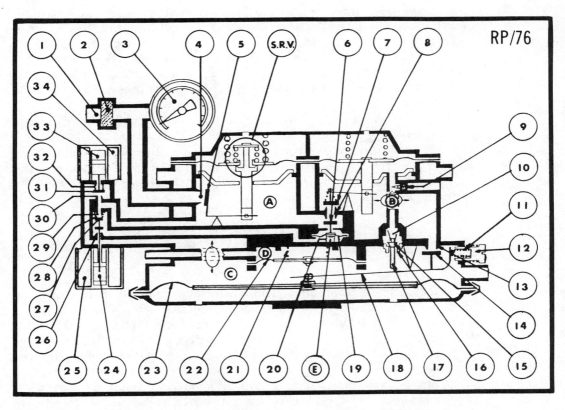

FIGURE 8-3. Tartarini CNG Pressure Regulator. *(Courtesy of Alternative Fuel Systems, Ltd.)*

speed. Fuel flow past seat 14 to the mixer will be proportional to engine rpm. The second stage supplies fuel at a constant pressure to the third stage and thus fuel pulsations in area A are prevented from reaching the final stage. Screw 12 adjusts idle mixture by changing the final-stage valve opening.

With the engine running, the manifold vacuum applied to area D holds diaphragm 22 in the upward position. When the engine is stopped, diaphragm 22 is forced down by the diaphragm spring. Downward movement of diaphragm 22 forces diaphragm 23 downward and closes seat 14, thereby providing positive fuel shutoff.

Fuelock 25 is deenergized when the engine is stopped, and valve 28 is then able to open. When valve 28 opens, fuel pressure from area A is applied to area E. Fuel pressure and spring tension move diaphragm 19 and seat 6 upward. The closing of seat 6 prevents any further fuel flow from area A. When the ignition switch is turned on, fuelock winding 26 is energized, seat 28 is closed, and first-stage fuel pressure is shut off from area E. Fuel pressure in area A will move diaphragm E and seat 6 downward and thus allow fuel to flow between the first and second stage. When coil 25 is energized to close seat 28, a small amount of fuel is vented from area E through seat 26 to area C to provide easier starting.

The energizing of prime solenoid 33 allows fuel to flow from area A to area C. Priming may be necessary for a few seconds on initial cold starting. Engine coolant is circulated through the pressure regulator to eliminate frosting.

With the engine idling, a clearance of 0.025 (0.635 mm) exists between lever 18 and power valve stem 17. At wide throttle opening, upward movement of lever 18 pushes power valve 17 open and allows additional fuel to flow from area B through the power valve to area C. The amount of mixture enrichment supplied by the power valve is preset by the power valve screw, item 9.

MIXERS The Tartarini dual fuel CNG mixer contains a venturi ring and movable venturi plate. CNG vapor is dispersed from a series of outlet holes around the venturi ring. In the CNG mode, the venturi plate rests against the venturi ring stop rivets. The diameter of the fuel discharge holes and the venturi plate height are matched to the engine CID. The mixer fits on top of the gasoline carburetor. In the gasoline mode, an electrically operated vacuum solenoid applies the vacuum to the mixer diaphragm to lift the venturi plate, as pictured in Figure 8-4. Mixers with fixed venturi plates are available for straight CNG conversions.

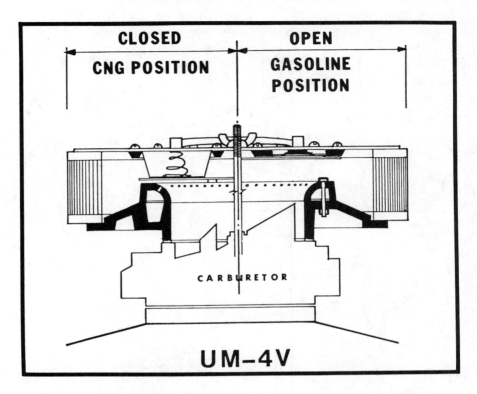

FIGURE 8-4. Tartarini CNG Mixer. *(Courtesy of Alternative Fuel Systems, Ltd.)*

Complete CNG Installations

1. *Filling connection.* Removal of the dust plug allows the probe refueling connection to be inserted in the fill fitting, as indicated in Figure 8-5. A one-way check valve automatically seals the system after refueling. The filling connection contains a burst disc.
2. *Cylinders.* CNG cylinders are selected from a variety of types and sizes to match the mounting space available in the vehicle. Most installations have two or three cylinders. Individual cylinder shutoff valves are required in some countries.
3. *Master shutoff valve.* The master shutoff valve allows fuel to be shut off at the cylinders if a leak occurs in the system. Normally, the master valve is left in the open position.

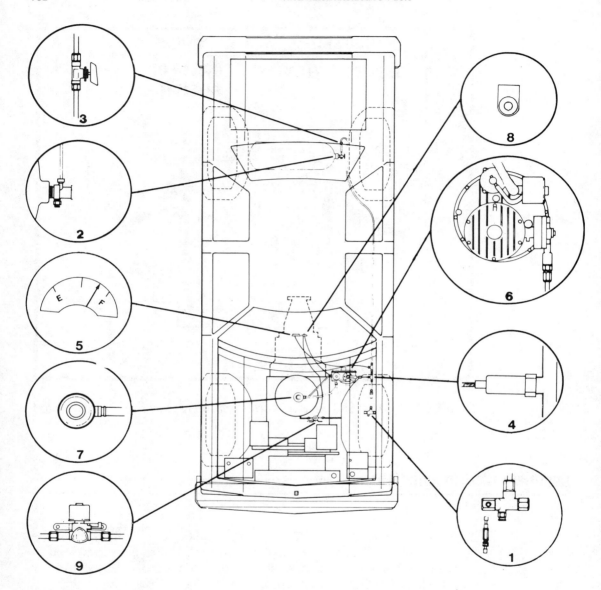

FIGURE 8-5. Complete CNG Fuel System. *(Courtesy of CNG Fuel Systems, Ltd.)*

4. *Fuel gauge transducer.* The transducer changes a pressure signal in the fuel line to an electrical signal to operate the fuel gauge.
5. *Fuel gauge.* The amount of CNG left in the cylinders is indicated on the gauge.
6. *Pressure regulator.* CNG at tank pressure is available at the regulator. The regulator reduces the fuel pressure and delivers the fuel to the mixer. Heater hose connections to the regulator would be similar to the connections for LPG conversions outlined in Chapter 4.
7. *Mixer or injector.* A vapor hose is connected from the regulator vapor outlet to the mixer. The mixer shown in Figure 8-5 is a venturi ring type specifically matched to the engine CID.
8. *Fuel selector switch.* The switch selects the fuel mode. Most CNG switches are the rotary type and are similar to the switches described in Chapter 4 for the LPG system.
9. *Gasoline fuelock.* The fuel selector switch energizes the gasoline fuelock when the driver selects the gasoline mode. When the gasoline fuelock is energized, gasoline is able to flow from the fuel pump to the carburetor.

Questions

1. A 12 3/4" × 4'9" CNG cylinder would contain the equivalent of _____ gallons of gasoline.
2. The fill connection is located between the first-stage regulator and the low-pressure regulator.   T F
3. Tank pressure is available in the first-stage regulator with the ignition switch in the off position.   T F
4. In a Tartarini CNG pressure regulator, the third stage is operated by _____ _____.
5. The Tartarini mixer venturi plate is lifted in the CNG mode.   T F
6. Richer air-fuel ratios at full throttle are provided by the _____ _____ in a Tartarini pressure regulator.
7. The first-stage and third-stage seats are closed with the ignition switch in the off position in a Tartarini pressure regulator.   T F
8. The CNG fuel gauge transducer changes a _____ signal to an _____ signal.

CHAPTER 9

Compressed Natural Gas— Regulations and Safety Precautions

Generally Accepted Regulations

CNG REGULATIONS There are many similarities between propane regulations and CNG regulations. In most countries, CNG cylinders cannot be manufactured or imported without government approval. The local authority in charge of high-pressure vessels is responsible for testing all CNG cylinders. State or provincial regulations regarding CNG conversions may vary to some extent. In some areas, CNG motor fuel is a new development and regulations may not be in place. We must emphasize that pressurized fuel technicians should familiarize themselves with all state or local regulations and adhere to them. We will discuss some generally accepted regulations.

CNG Cylinders

1. CNG fuel cylinders must have a design pressure of not less than 3,000 psi (204 atm).
2. CNG cylinders must meet Department of Transport (DOT) regulations.
3. Welding to any part of a fuel cylinder, except brackets or saddle plates, requires approval by the pressure vessel inspection authority.

4. Each cylinder must be equipped with a fusible burst disc set to have a fusible backing yield of 212°F (100°C) and a burst rating of 4,200–5,000 psi (286–345 atm).
5. Components directly in contact with fuel cylinders must be electrochemically compatible with the cylinder. For example, brass shall not come in contact with aluminum.

CNG Cylinder Mounting

1. Fuel cylinders must not be installed on the roof of a vehicle.
2. Fuel cylinders must not be installed in the engine compartment.
3. CNG cylinders must not project beyond the sides of a vehicle.
4. CNG cylinders must not be mounted ahead of the front axle.
5. CNG cylinders must not project above the highest part of the vehicle.
6. The minimum clearance between the fuel cylinder and the exhaust system is 2 in (5 cm).
7. When a fuel cylinder is mounted between the vehicle axles, clearance between the cylinder and the road surface shall be 7 in (17.5 cm) for vehicles under a 127-in (318-cm) wheel base. Vehicles with a wheel base in excess of 127 in (318 cm) shall have a clearance of 9 in (24 cm) between the fuel cylinder and the road surface.
8. If the fuel cylinders are mounted behind the rear axle and below the frame, the clearance between the fuel cylinder and the road surface shall be 8 in (20 cm). On vehicles having more than 45 in (112 cm) between the rear axle and fuel cylinder center lines, fuel cylinder road clearance shall be 0.18 times the distance between the center lines of the rear axle and fuel cylinder.
9. All CNG cylinder valves must be protected from possible damage.
10. Mounting bands for fuel cylinders under 100 l capacity shall be 1.2 in (30 mm) wide and shall provide strength equivalent to that of steel 140 in^2 (896 cm^2) in cross-sectional steel area. Band bolt holes shall not exceed 0.430 in (11 mm) in diameter. Attaching bolts shall be a minimum of 3/8 in (10 mm) in diameter.
11. Mounting bands for fuel cylinders over 100 l capacity shall be 1.8 in (45 mm) wide and shall provide strength equivalent to that of steel 350 in^2 (2,240 cm^2) in cross-sectional area. Band bolt holes shall not exceed 0.500 in (12 mm) in diameter. Attaching bolts shall be a minimum of 0.500 in (12 mm) in diameter.
12. Fuel cylinders must be mounted so the vehicle structure is not significantly weakened. Wherever necessary, reinforcement shall be

added to the vehicle structure so the force necessary to separate the cylinder from the vehicle is twenty times the mass of a full cylinder in the longitudinal direction, and eight times the mass of a full cylinder in any other direction.

13. Fuel cylinders shall not cause a vehicle to exceed gross vehicle weight limitations. Fuel cylinder weight must not create any unsafe load distribution on the vehicle axles or tires.
14. When fuel cylinders are mounted in a vehicle compartment that could be used to transport people, the neck of the cylinder and fittings shall be enclosed in a gas-tight enclosure that is vented outside the vehicle.
15. When fuel cylinders are mounted in a vehicle compartment that could not be used to transport people, the cylinders shall be mounted as described in regulation 14, or the mounting compartment shall be ventilated with an opening of 3 in^2 (19.2 cm^2) at the highest practical point.

Piping, Tubing Requirements

1. Piping located upstream of the first-stage regulator shall withstand a pressure of four times the maximum cylinder working pressure.
2. Piping located downstream of the first-stage regulator shall withstand a pressure of five times the maximum working pressure.
3. Copper tubing must not be used as a fuel supply line.
4. Piping, tubing, fittings, and hose shall be constructed of a material resistant to the action of natural gas.
5. Piping tubing or hose shall have sufficient internal size to provide the required flow of fuel.
6. When piping, tubing, or hose is mounted, provision shall be made for vibration or movement of equipment.
7. All piping shall be supported at intervals not exceeding 2 ft (61 cm) with plastic-coated metal clamps. Some areas may allow nylon ties for fuel line clamps.
8. Piping joints shall be threaded.
9. A piping or fitting thread shall be tapered and comply with ANSI Standard B2.1 "Pipe Threads (except dryseal)."
10. A piping or fitting thread subjected to container pressures shall comply with ANSI Standard B2.2 "Dryseal Pipe Threads." Container threads shall comply with ANSI Standard B57.1 "Compressed Gas Cylinder Valve Outlet and Inlet Connections."

11. Jointing material must be an approved type and shall be applied to the male pipe threads only.
12. Seamless tubing joints subjected to cylinder pressure shall be a double inverted-flare joint complying with SAE J533 "Flares for Tubing."
13. All joints and connections must be accessible.
14. Concealed piping must not be located where undetected leakage could cause an accumulation of gas.
15. Fittings containing both left- and right-hand threads shall not be used.
16. Bends in piping must not weaken or reduce the internal area of the pipe.
17. A bushing other than steel or brass must not be used.
18. Structural vehicle members must not be cut in such a manner as to reduce the strength of the member.
19. A quick-disconnect coupling shall not be substituted for a manual shutoff valve.
20. Defective fuel lines must be replaced; repairing of fuel lines is prohibited.
21. Fuel lines must not be located in the drive shaft tunnel.
22. Fuel lines must be located in the most practical position that would afford protection from an impact or collision.
23. A grommet must be used to protect fuel lines passing through metal panels in the vehicle.
24. Metal shielding must be located within 1 in (2.5 cm) of the fuel line when the fuel line is located within 8 in (20 cm) of the exhaust manifold or 4 in (10 cm) of any other component of the exhaust system.
25. All fuel lines must be pressure tested using gas, air, or inert gas such as carbon dioxide. The fuel line shall retain the maximum working pressure for at least 10 minutes without showing any drop in pressure. The source of test pressure shall be isolated before the reading begins. Pressure must be measured with a pressure gauge or equivalent device. When a leak is indicated by the test, the source of the leak must be located with an approved leak detector solution or an electronic tester. The leak must be corrected and the pressure test repeated.
26. All fittings must be tested for leaks after the conversion is completed.
27. Leak testing must be done in a ventilated area located away from all sources of ignition.

28. When a vehicle is involved in a collision involving any part of the CNG system, all fuel lines and fittings must be leak tested.

Valve Requirements

1. A manual shutoff valve must be installed in a location to permit isolation of the fuel cylinders from the rest of the CNG system.
2. An automatic valve must be installed downstream of the manual shutoff valve to prevent fuel from flowing to the mixer if the engine ceases to rotate, or if the ignition is shut off.
3. The refueling connection must be equipped with a back-check valve to prevent the return flow of fuel from the fuel cylinders to the filling connection.
4. The vehicle refueling intake shall be provided with a burst disc located between the back-check valve and the refueling intake. This disc does not require fusible backing, and the burst pressure rating is 4,200–5,000 psi (286–345 atm).

Gauge Requirements

1. Every CNG fuel system must be equipped with a pressure gauge.
2. Pressure gauges mounted in the passenger compartment must be a type of gauge that eliminates the use of gas lines in the passenger compartment.

CNG Safety Precautions

The Italian experience has proved that CNG vehicles can have an excellent safety record if regulations and safety precautions are followed. Listed below are some of the safety precautions that should be followed when CNG-equipped vehicles are being serviced.

1. If natural gas escapes in a closed shop, shut off all sources of ignition, such as pilot lights, and ventilate the area. Natural gas tends to rise in the atmosphere when it escapes. Be certain that overhead open flame heaters are shut off completely if gas escapes.
2. Locate CNG vehicle service areas away from overhead heaters and welding equipment.

3. Shut off the manual fuel valve and run the engine until it stops before disconnecting the fuel lines between the manual valve and the regulators.
4. All fuel line connections must be tested for leaks on completion of a CNG conversion. Liquid leak-detecting solutions or an electronic tester may be used.
5. If natural gas odor occurs in a vehicle, all lines, fittings, and pressure regulators should be tested for leaks.
6. Periodically, all fuel lines should be visually inspected and all fuel line fittings should be leak tested.
7. Some local regulations require fuel cylinder shutoff valves to be closed when CNG-powered vehicles are stored inside.
8. Do not smoke when servicing or refueling CNG-equipped vehicles.
9. Obey all regional conversion and fire regulations.

―――――――――――――――― Questions ――――――――――――――――

1. CNG fuel cylinders have a burst disc with a fusible backing yield of _____ degrees F.
2. A brass fitting in an aluminum fuel cylinder would be acceptable.
   T   F
3. The minimum clearance between fuel cylinders and the exhaust system is _____.
4. Copper tubing may be used for CNG fuel lines.   T   F
5. CNG fuel lines must be heat shielded if they are located within 8 in or less of the exhaust manifolds.   T   F
6. The burst disc in the filling connection requires a fusible backing.
   T   F
7. Gas lines to operate a pressure gauge may be located in the passenger compartment.   T   F

CHAPTER 10

Filling Facilities For Compressed Natural Gas

Compressors

RECIPROCATING COMPRESSORS The reciprocating compressor illustrated in Figure 10-1 is an electrically powered four-stage unit. Natural gas enters the first stage from the city gas main. Gas compressed in the first stage flows into the second stage, where it is further compressed. Natural gas flows through each stage in succession, and each stage boosts the gas pressure. One-way check valves prevent reverse flow of gas between stages. Intercoolers and condensate traps may be used between stages. The compressor pictured in Figure 10-1 is air cooled; other reciprocating compressors may be water cooled.

A preset pressure switch starts and stops the compressor automatically. When the compressor stops, all pressure in the cylinder intercoolers and aftercoolers is automatically relieved. Condensate from the traps is drained to an exterior tank when the compressor stops. Compressors will operate more efficiently at higher inlet pressures. Minimum hourly inlet flow would be 1,800 ft^3/min at 5 psi (35 kp). Higher inlet pressures reduce the number of compressor stages. A two-stage compressor would be adequate with an inlet pressure of 150 psi (1,050 kp). Compressors are available with different output ratings from 1 ft^3/min (cfm) to 300 cfm. A slow-fill compressor rated at 2 cfm would fill two vehicles in a 10-hour period. Fast fill compressors with output ratings above 75 cfm are designed to continuously fill vehicle CNG cylinders in a few minutes. Most compressors are connected so they can fill a bank of fuel cylinders referred to as a cascade. Vehicles can be filled quickly from

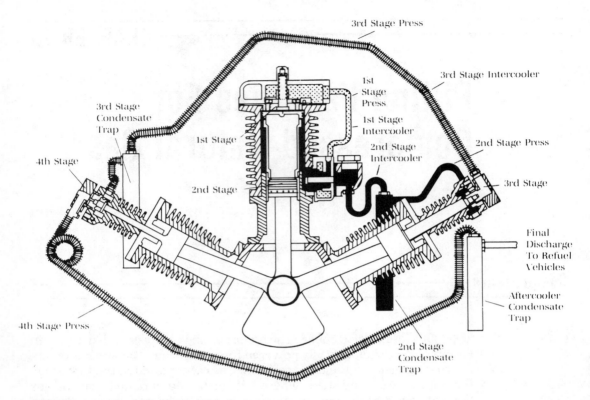

FIGURE 10-1. Reciprocating Compressor. *(Courtesy of Dual Fuel Systems, Inc.)*

the cascade. For smaller fleets, a lower output compressor of about 30 cfm output might be used to fill the cascade. (Compressor and cascade connections are illustrated in Figure 7-6.)

HYDRAULIC COMPRESSORS An electric motor is used to drive an oil pump in the hydraulic compressor, as illustrated in Figure 10-2.

Oil from the pump enters area H1 and H2 in Figure 10-3. Intake gas enters the first-stage combustion chamber (area 1) through the intake check valve. Hydraulic pressure in fluid chamber H1 drives the piston assembly to the left and thus compresses the gas in area 1. Compressed gas from area 1 flows through cooling coil A to the second stage (area 2). Hydraulic pressure in fluid chamber H2 drives the piston assembly to the right, thereby compressing the gas in (area 2) and drawing a fresh charge of gas into area 1. Gas flows from area 2 through cooling coil B to the third stage (area 3). Hydraulic pressure in fluid chamber H1 drives the piston assembly to the left and thus compresses the gas in areas 1 and 3. Highly compressed gas in area 3 flows into the fourth stage (area 4). Hydraulic pressure in fluid chamber H2 drives the piston assembly to the right and compresses the gas in areas 2 and 4. Highly compressed gas from area 4 flows through cooling coil D to a cascade of fuel cylinders. Vehicles are refueled from the cascade.

Electronic proximity switches at each end of the compressor sense the position of the piston assembly. High-pressure oil is directed alternately to areas H1 and H2 by a pair of valves operated by the electronic proximity switches. Refueling cascades are usually filled to 3,600 psi (245 atm). Vehicle refueling pressure is 2,400–3,000 psi (163–200 atm). A slow-fill compressor that would refuel one or two vehicles in a 10-hour period would cost $4,000–5,000. The cost of a fast-fill, high-cfm-rated compressor would be approximately $80,000. Local fire, electrical, and CNG refueling station regulations have to be followed in the construction of a CNG filling station.

FIGURE 10-2. **Hydraulic Compressor.** *(Courtesy of CNG Fuel Systems, Ltd.)*

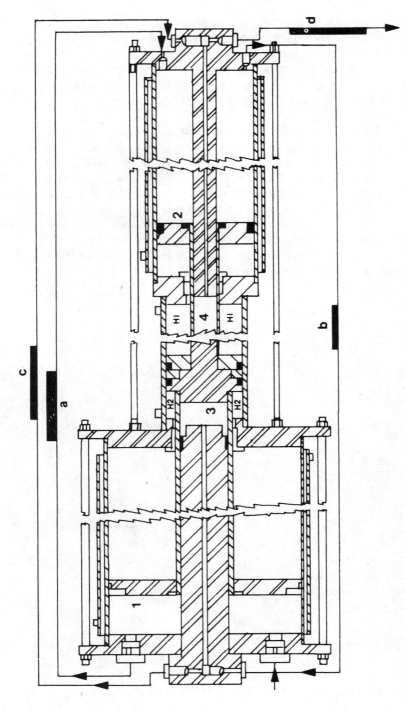

FIGURE 10-3. Internal Components of Hydraulic Compressor. *(Courtesy of CNG Fuel Systems, Ltd.)*

174

CNG Measuring Devices

U-TUBE MAGNETIC SENSING METERS One type of CNG flow meter employs a stainless steel U-tube. The gas or fluid being measured is forced through the unobstructed U-tube. Gas velocity causes the U-tube to twist, as pictured in Figure 10-4. Two magnetic position detectors, one on each side of the U-tube, generate signals in relation to movement of the U-tube. Solid state circuitry in the flow meter changes the position detector signals to digital or meter readings, as indicated in Figure 10-5.

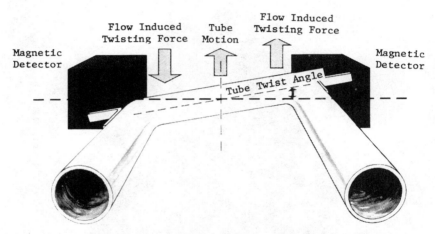

FIGURE 10-4. Flow Meter U-tube Principle. *(Courtesy of Micro Motion, Inc.)*

Questions
---

1. A gas compressor operating at 150 psi (1,050 kp) inlet pressure would require more stages than a compressor operating at 10 psi (70 kp) inlet pressure.　　　　　　　　　　　　　　　　　　　　　T　F

2. In a hydraulic gas compressor, the pumping piston is moved by _____ _____.

3. Cascades in a CNG filling station are usually filled to a pressure of _____.

4. In most fast-fill refueling stations, vehicle cylinders are filled directly from the compressor.　　　　　　　　　　　　　　　　　　T　F

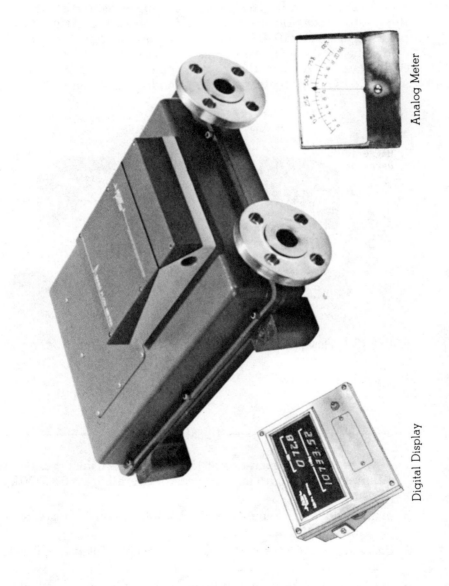

FIGURE 10-5. CNG Flow Meter. *(Courtesy of Micro Motion, Inc.)*

CHAPTER 11

Liquid Natural Gas

General Facts

NATURAL GAS LIQUEFACTION PROCEDURE The basic operations in liquefying natural gas are: gas treatment, liquefication, and storage. As mentioned in Chapter 2, the boiling point of methane or natural gas is $-260°F(-162°C)$. Natural gas is liquefied by lowering the temperature to the boiling point. Before the gas can be liquefied, contaminants that would solidify at extremely cold temperatures must be removed. Gas treatment involves the removal of contaminants such as water, carbon dioxide, sulphur compounds, dust, oil, and heavy hydrocarbons. A contaminant removal system consists of an inlet separator, molecular sieves to remove moisture and carbon dioxide, and a compressor.

The liquefaction unit consists of a cold box, refrigerant compressor, coolers, and refrigerant storage. Storage containers for liquid natural gas (LNG) contain special insulation to maintain the low temperature of the liquid. Fuel tank materials have to be compatible with LNG. Storage tanks vary widely in size and structure.

The use of LNG has steadily increased in the past two decades. Many gas utility companies liquefy and store natural gas to meet peak demands in extremely cold weather. During peak demand periods when natural gas pipeline capacity may not be able to meet the demand, LNG is vaporized and released into the pipelines. The storage of LNG to meet peak demand periods is referred to as peak shaving. The Northern Indiana Public Service Company has an LNG peak shaving plant at La Porte, Indiana. Tank storage capacity is 580,000 bbl (96,700 m^3), which is equivalent to 2 billion ft^3 (57 million m^3) of natural gas.

BTU CONTENT AND COST FACTORS OF LNG Liquid natural gas contains 83,000 BTUs per U.S. gal. LNG has an octane rating of 130 and weighs 3.43 lb per U.S. gal. A natural gas liquefaction plant with a 9,000 gal (40,500 l) per day capacity would cost approximately $2,500,000. The cost of an LNG refueling facility would be around $250,000. LNG would cost approximately 53¢ per gallon at privately owned filling facilities, or 65¢ per gallon at public refueling stations. Approximate LNG conversion costs would be $2,200 per vehicle. Conversion costs will vary widely, depending on fuel tank size, type of conversion equipment, and labor rates.

LNG FUEL TANKS The liquid methane fuel tank pictured in Figure 11-1 has a stainless steel inner tank and a carbon steel outer shell. The space between the inner and outer shell is insulated with a special aluminized mylar material and maintains a permanent high vacuum to hold the liquid fuel at $-260°F$ ($-162°C$).

The inner stainless steel tank is a beam-supported pressure vessel. A liquid-level sensing unit in the tank indicates the fuel level on a dash-mounted gauge. Flow control valves and pressure relief valves are mounted in the tank. Tank operating pressure would be 20–60 psi (140–420 kp). Heat loss on an LNG fuel tank would be approximately 4.69 BTUs per hour. As heat is lost from the tank, the liquid fuel slowly changes to a vapor, and tank pressure increases. The tank lockup time is the time required to pressurize the tank without venting. During the lockup time,

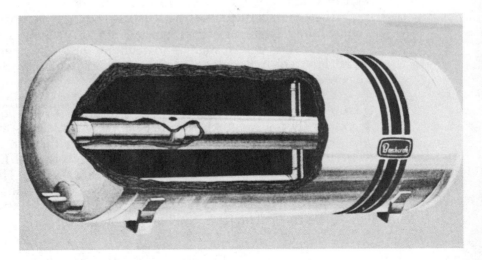

FIGURE 11-1. Liquid Methane Fuel Tank. *(Courtesy of Beech Aircraft Corporation, Alternative Energy Division)*

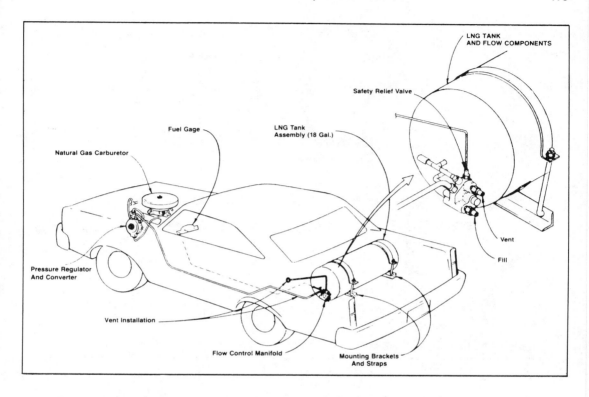

FIGURE 11-2. LNG Fuel Tank Installation. *(Courtesy of Beech Aircraft Corporation, Alternative Energy Division)*

tank pressure will increase from 20 psi (140 kp) to 60 psi (420 kp). The relief valve will vent the tank momentarily at 60 psi (420 kp). Tank lockup time is over 7 days, and boil-off losses following lockup time are less than 1 percent per day. Under normal driving conditions, a vehicle would not be left sitting long enough for tank lockup time to be completed. The tank pictured in Figure 11-2 holds 18.4 gallons of LNG and weighs 108 lb. (48.6 kg). Larger tanks are available. As a motor fuel, LNG solves the problems of weight and the lack of driving range that arise with CNG fuel cylinders. The relief valve illustrated in Figure 11-2 is vented outside the vehicle.

The fuel tank must be vented during the filling operation. Tank fill and vent hoses are connected to the fuel tank fittings with quick couplers, as indicated in Figure 11-3. Gas vented from the tank during refueling may be routed back into city gas mains. Vented gas may be stored in a separate storage vessel and burned off at a constant rate.

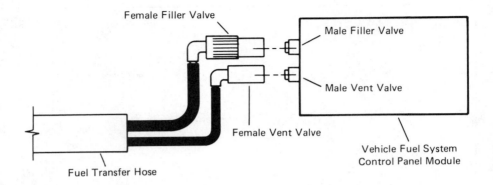

FIGURE 11-3. Tank Fill and Vent Quick Couplers. *(Courtesy of Essex Cryogenics of Missouri, Inc.)*

LNG Conversion Systems

LIQUID MODE OPERATION In the liquid operating mode, fuel is forced from the fuel tank by tank pressure. Liquid natural gas flows through the shutoff valve to the flow control valve. The flow control valve controls the operating mode. When tank pressure is below 40 psi (280 kp), the flow control valve remains in the liquid mode. Tank pressure above 40 psi (280 kp) switches the flow control valve to the vapor mode. A trunk-mounted LNG tank is illustrated in Figure 11-4. The flow control valve is mounted on the control panel beside the tank.

Liquid fuel flows from the flow control valve to the solenoid valve and combination pressure regulator and heat exchanger. The LNG solenoid valve is very similar to the propane fuelock described in Chapter 4. Engine coolant is circulated through the heat exchanger by heater hoses connected to the cooling system. Heat from the cooling system assists in vaporizing the liquid fuel in the heat exchanger. The combination pressure regulator and heat exchanger would be similar to the LPG vaporizers discussed in Chapter 3. Fuel vapor flows from the pressure regulator-heat exchanger to the air-fuel mixer. The air-fuel mixer is basically the same as the mixers used for LPG or CNG. The mixer may be adapted to the top of the gasoline carburetor for dual fuel operation. On straight LNG systems, the mixer could be adapted to the

FIGURE 11-4. LNG Tank and Control Panel. *(Courtesy of Beech Aircraft Corporation, Alternative Energy Division)*

base of the gasoline carburetor. The LNG motor fuel industry is in its infancy. Because of the lack of refueling facilities, most LNG systems will be dual fuel. Figure 11-5 shows an LNG system in the liquid mode.

VAPOR MODE OPERATION The flow control valve switches to the vapor mode when tank pressure exceeds 40 psi (280 kp). Vapor mode operation would occur after the vehicle had not been used for a few days, especially in hot weather. Fuel vapor flows from the top of the fuel tank through the one-way check valve and flow control valve. The one-way check valve allows vapor to flow from the tank to the flow control valve, but prevents any reverse flow of liquid fuel from the control valve to the relief valve during refueling. Vapor flows from the flow control valve through the LNG solenoid, and the pressure regulator–heat exchanger to the air-fuel mixer. Operating in the vapor mode will reduce tank pressure, and the flow control valve will switch back to the liquid mode. Vapor mode fuel flow is depicted in Figure 11-6.

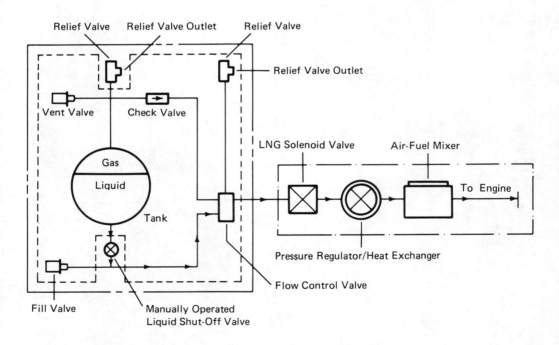

FIGURE 11-5. LNG Liquid Mode Operation. *(Courtesy of Essex Cryogenics of Missouri, Inc.)*

Liquid Natural Gas   183

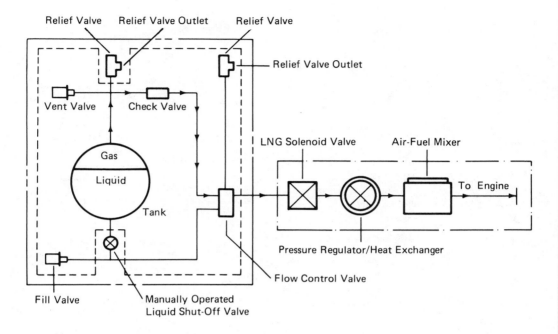

FIGURE 11-6. LNG System Vapor Mode Operation. *(Courtesy of Essex Cryogenics of Missouri, Inc.)*

LNG ELECTRICAL SYSTEM A vacuum switch containing a set of normally open (NO) contacts, and a set of normally closed (NC) contacts, is connected between the ignition switch and the fuel solenoid valves. While the engine is cranking, current flows from the vacuum switch through a fuel selector switch to one of the fuel solenoids, depending on the fuel selector switch position. Once the engine starts, the manifold vacuum will close the NO vacuum switch contacts. Current will flow from the ignition switch run terminal through the vacuum switch NO contacts and fuel selector switch to the fuel solenoid. The LNG system wiring diagram is illustrated in Figure 11-7.

The full level switch and the fuel quantity sender are located in the fuel tank. The control panel module is located near the fuel tank. Instrument panel gauge operation or control panel gauge operation is selected by the control panel selector switch. When the control panel selector switch contacts are connected between the center and lower

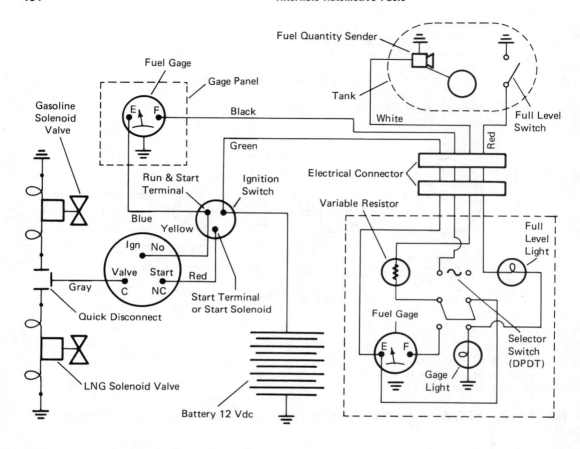

FIGURE 11-7. LNG System Wiring Diagram. *(Courtesy of Essex Cryogenics of Missouri, Inc.)*

switch terminals, the control panel gauge will be in operation. Current will flow from the ignition switch through the control panel fuel gauge to the variable resistor and the fuel quantity sender. When the tank is being refueled, the full level switch will close at the maximum fuel level. Closing of the full level switch contacts completes the circuit from the empty side of the control panel fuel gauge through the full level light to ground. Illumination of the full level light indicates completion of the filling operation. When the control panel switch contacts are connected between the center and upper switch terminals, the instrument panel fuel gauge is connected through the selector switch contacts and variable resistor to the fuel quantity sender. The variable resistor provides for fuel gauge adjustment.

ENGINE TUNING REQUIREMENTS Ignition and air-fuel ratio requirements for LNG fuel systems are very similar to the requirements for a CNG-fueled engine. In either fuel system, natural gas vapor is available at the mixer.

LNG Refueling Facilities

PRESSURE TRANSFER FILLING STATIONS The bulk storage tank, fuel transfer system, vehicle couplings, and vapor disposal system are the main components in any LNG refueling facility. Pressure transfer stations use storage tank pressure to refuel vehicle tanks. System operating pressure is controlled by a heat exchanger and pressure-regulating valve. The pressure transfer systems operate without the use of external power sources such as electric motors. In remote areas, where electricity is not readily available, a pressure transfer refueling facility could be employed. The disadvantage of the pressure transfer system is the difficulty that arises in disposing of gas vented from the fuel tank during refueling. In other types of filling facilities, vented gas is returned to the LNG storage tank. Vented gas cannot be returned to the storage tank in a pressure transfer system because the pressure difference between the storage tank and the vehicle fuel tank is too small. Vented gas must be routed to a gas distribution line or to a storage tank. Stored gas may be used to heat buildings. Figure 11-8 shows a filling station that would meet the requirements of a small fleet.

IN-LINE PUMP FILLING STATION An electrically driven pump delivers fuel to the vehicle fuel tank in an in-line pump refueling station. Tank pressure forces the fuel from the tank to the pump. The external pump must be cooled prior to operation to prevent fuel vaporization. LNG from the storage tank is circulated around the pump, for cooling purposes, and routed back to the storage tank. Heat given off by the pump is transferred to the LNG and increases storage tank pressure. During the refueling operation, vapor from the vehicle fuel tank is returned to the bottom of the storage tank. As the vapor rises through the LNG, some of the vapor will condense and the remaining vapor will be cooled, thereby minimizing the increase in storage tank pressure. A large commercial LNG refueling facility is pictured in Figure 11-9.

SUBMERGED PUMP FILLING STATIONS The LNG pump and drive motor are mounted in the storage tank in a submerged pump system. Pump cooling

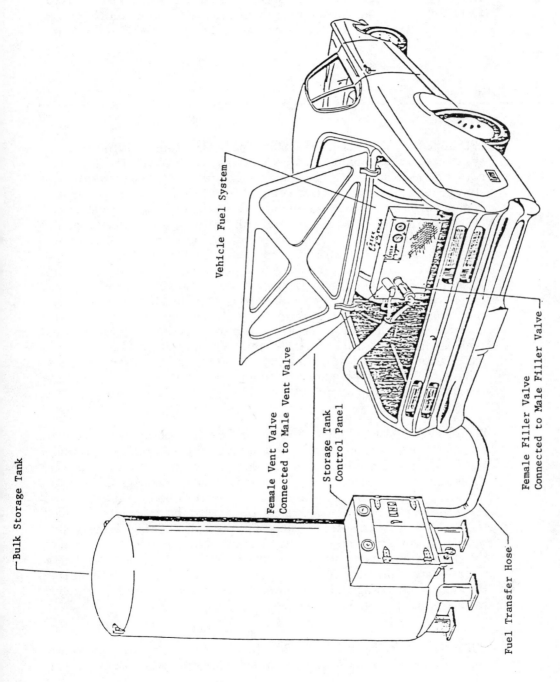

FIGURE 11-8. LNG Small Fleet Refueling Facility. *(Courtesy of Essex Cryogenics of Missouri, Inc.)*

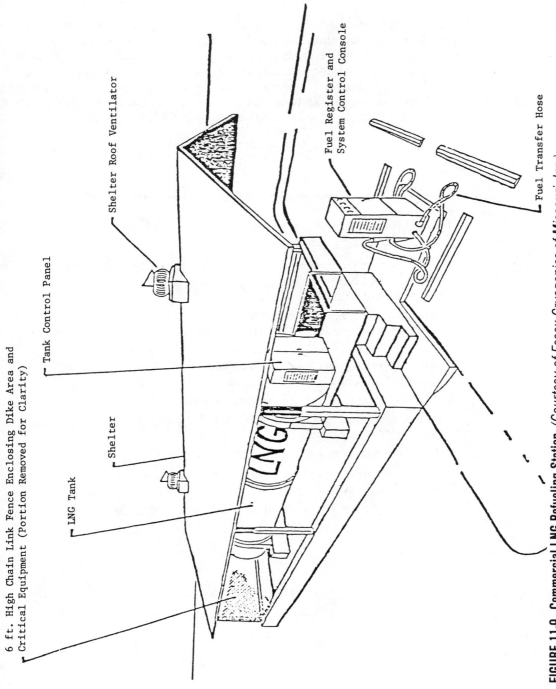

FIGURE 11-9. Commercial LNG Refueling Station. *(Courtesy of Essex Cryogenics of Missouri, Inc.)*

is provided by the LNG. Pump pressure forces the fuel from the storage tank to the vehicle tank. Removable portholes in the storage tank are provided for pump replacement. Pump service would be complicated because of the necessity of draining the storage tank. A submerged pump could not be used in a small storage tank. Another drawback is that accidental spillage of LNG will cause severe frostbite. Gloves and face masks should be worn when LNG-powered vehicles are being refueled.

--- Questions ---

1. Liquefication of natural gas vapor requires cooling to _____ _____ degrees F.
2. LNG contains _____ BTUs per U.S. gallon.
3. The cost of LNG would be _____ cents per gallon from a public refueling station.
4. The space between the two sections of an LNG fuel tank is insulated with an aluminized mylar material and maintained in a permanent _____ _____.
5. The operating pressure in an LNG fuel tank would be from _____ to _____ psi.
6. LNG fuel tanks are vented during refueling.   T   F
7. When LNG fuel tank pressure is 50 psi, the fuel system will be in the liquid operating mode.   T   F
8. Liquid or vapor operating mode is determined by the _____ _____ _____.
9. The full level switch operates the control panel fuel gauge.   T   F
10. Pump cooling is required on an in-line pump refueling station.   T   F

CHAPTER 12

Alcohols as Motor Fuels

Ethanol

CHEMICAL COMPOSITION AND BTU CONTENT The chemical symbol for ethanol is C_2H_5OH. Ethanol differs from hydrocarbon fuels because it contains hydroxy radical (OH) as well as carbon and hydrogen. Ethanol contains approximately 35 percent oxygen by weight. The BTU content of ethanol is 76,000 per gal. Alcohols have a much greater cooling effect than gasoline when they are changed from a liquid to a vapor in the carburetor. Latent heat of vaporization is an indication of the cooling effect when liquids are vaporized. Gasoline has a latent heat of vaporization of 800 BTUs per gal compared with ethanol, which has 2,600 BTUs per gal. Ethanol has more than three times as much cooling effect as gasoline when it is vaporized. Increased cooling effect provides a denser air-fuel mixture in the intake manifold and engine cylinders. The higher latent heat of vaporization and denser air-fuel mixture compensate for the reduced BTU content of ethanol compared with gasoline. Engine power with ethanol and gasoline mixes should be similar to power output with gasoline.

FEEDSTOCK Ethanol can be manufactured from agricultural products containing carbohydrates. Corn, barley, wheat, potatoes, sugar beets, and sugarcane could be used as ethanol feedstocks. Most raw feedstocks are converted to starch and the starch changed into sugar. Ethanol is derived from sugar by a distillation process. If sugar beets or sugarcane are used as feedstocks, the starch does not have to be converted to sugar in the manufacturing process. In the United States the most common ethanol feedstock is corn. Brazil manufactures large quantities of ethanol from sugarcane for use as a motor fuel.

PROCESSING PLANTS AND COST FACTORS Ethanol plants may vary in size from small farm operations producing 40,000 gal per year to large commercial operations producing millions of gallons per year. Many small personally owned ethanol plants exist in the United States at the present time. Commercial ethanol plants have a low output compared with gasoline refineries because alcohols have not been widely accepted as motor fuels in the United States. A small commercial ethanol plant might produce 1.5 million gal per year. Ethanol has usually been marketed commercially as a gasoline extender, in a mix of 10 percent ethanol with gasoline. The mixture of gasoline and ethanol is referred to as gasohol. The floor plan of a small farm ethanol plant is illustrated in Figure 12-1. The basic steps in ethanol production are outlined in the following paragraphs.

Day 1

The tank is filled with 900 gal of 190°F (88°C) water. Fifty bushels of ground corn are added slowly to the water to prevent lumping problems. The pH level of the batch is checked and corrected as necessary. Process chemicals are added to break down the starch in the corn and produce dextrins. Gelatinization, or thickening, of the mash occurs as the starch is liquefied. The tank is held at 185°F (85°C) for 8-12 hours.

Day 2

The tank is cooled to 140°F (60°C). Temperature reduction requires 8-9 hours. Dissipated heat is used to assist in warming incoming water for a new batch in another tank. The pH or acid level is adjusted when the temperature reaches 140°F (60°C). Bacteria protection is provided by adding sodium bisulfite. Tank temperature is maintained at 140°F (60°C) for 8-12 hours.

Day 3

Sugar content is tested and tank temperature is adjusted to 90°F (32°C). When the batch is cooled, the pH is adjusted and yeast is added.

Days 4 and 5

Sixty hours are required for the fermentation stage when live yeast cells

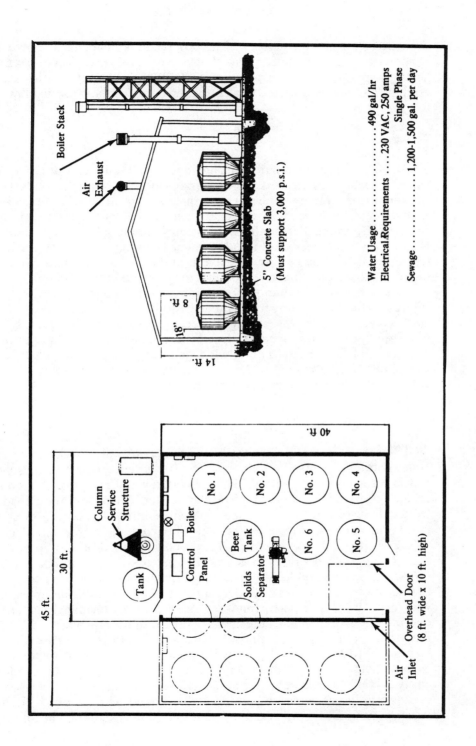

FIGURE 12-1. Ethanol Plant Floor Plan. *(Courtesy of Harvest Fuel, Ltd.)*

are converting sugar to alcohol. The process tank develops a blanket of carbon dioxide over the contents that prevents the intrusion of oxygen. Tank temperature is maintained at 90°F (32°C). When the entire sugar content has been converted to alcohol, the batch is ready for distillation.

Day 6

The mash is pumped from the tank into a two-part solid separator. Liquid, called beer, is pumped into a storage tank prior to distillation. Solids from the separator tank are referred to as distiller's grain, which is used as a high-protein cattle feed. A screw auger lowers moisture content in the solids to 60 percent. Fermented beer is approximately 12 percent alcohol. Distillation of the beer produces 115 gal of 190 proof alcohol. Waste heat from the distillation column is used to preheat water for a new batch. Figure 12-2 pictures the inside of a farm-operated ethanol plant.

Cost factors for ethanol production are outlined in Table 12-1. Cost figures on the left assume the plant operator will use propane boiler fuel and electricity as process energy. The cost estimates on the right assume the plant operator has access to free process energy. Waste agricultural products for boiler fuel and a wind generator to produce electricity would eliminate the cost of process energy. Ethanol production costs vary widely, depending on the feedstock cost.

GASOHOL As mentioned previously, ethanol is usually marketed in the United States as a gasoline extender. The 10 percent ethanol and gasoline mix is referred to as gasohol. Ethanol with a proof rating of 190 would contain 95 percent alcohol and 5 percent water. The proof rating divided by two determines the percentage of alcohol content. If the water content in ethanol exceeds certain limits in gasohol fuel, the water and alcohol will separate from the gasoline. This process is called phase separation. An engine will operate satisfactorily on gasohol if minor tuning modifications are made. Major carburetor modifications are necessary to operate the engine on high percentages of alcohol. Phase separation will result in severe performance problems in an engine designed to operate on gasohol or gasoline because pure ethanol and water will enter the fuel system. To prevent phase separation, alcohol proof rating should be as close to 200 as possible. Cosolvents, such as n-butanol, mixed with gasohol help to prevent phase separation. Phase separation occurs more easily at cold temperatures. Ethanol has an octane rating of 102. Gasoline octane rating is boosted when alcohol is mixed with gasoline. Higher compression ratios could be used with alcohols or gasohol. The amount of compression ratio increase would depend on the percentage of alcohol being used.

TABLE 12-1. Ethanol Production Cost Factors

	Cost per gallon of alcohol	Special case costs gal/alcohol
Feedstock cost (Corn @ $2.50/bu. @ 2.3 gal/bu)	$1.09	$ .87 ($2/bu)
Boiler fuel ($.60/gal. propane)	$ .23	-0-
Process chemicals (enzymes, acids, bases, disinfectants, etc.)	$ .24	$ .24
Labor ($5/hr. 4-hrs daily/100 gal)	$ .20	$ .20
Electricity ($ .05/kwh)	$ .10	-0-
sub-total	$1.86	$1.31
Retained value of feedstock (70% of corn price fed to dairy cows)	−$ .76	−$ .61
sub-total	$1.10	$ .70
User tax credit (USA ONLY). (for ethanol between 150-190 proof)	−$ .30	−$ .30
TOTAL COST	$ .80	$ .40

SOURCE: Courtesy Harvest Fuel Ltd.

FIGURE 12-2. Ethanol Plant Internal Design. *(Courtesy of Harvest Fuel, Ltd.)*

Methanol

CHEMICAL COMPOSITION AND BTU CONTENT The chemical symbol for methanol is CH_3OH. The oxygen content of methanol is about 50 percent by weight. Hydroxy radical (OH) is also present in methanol. Methanol contains 57,000 BTUs per gal. The latent heat of vaporization for methanol is 3,300 BTUs per gal. Compared with gasoline, methanol has approximately four times as much cooling effect when it changes from a liquid to a vapor in the carburetor. The increased cooling effect provides a denser air-fuel mixture and increases engine power. Methanol air-fuel ratios must be much richer than gasoline air-fuel ratios. Air-fuel ratios for alcohols are discussed in the following chapter on engine tuning and emission levels for alcohol fuels. Methanol has been a favorite fuel for many race car drivers because of the latent heat of vaporization.

FEEDSTOCK Methanol can be manufactured from lignite or coal, municipal waste, agricultural waste, or forestry waste. It is estimated that the United States has 320 million tons of agricultural waste and 120 million tons of forestry waste per year. If all these waste products were converted to methanol, U.S. crude oil consumption would be reduced by one-third. A city having a population of 3 million would create enough waste to produce 11,000 bbl of methanol per day. The same city would require 80,000 bbl of motor fuel per day.

PROCESSING PLANTS AND COST FACTORS Methanol produced from lignite or coal would cost approximately 35¢ per gal. The cost of producing methanol in a waste plant would be about 50¢ per gal. Production costs vary, depending on feedstock and transportation costs. If alcohols were used as motor fuel by a significant percentage of the U.S. vehicle fleet, a whole new alcohol production and distribution system would be necessary. U.S. methanol production is about 85,000 bbl per day, while motor fuel requirements are approximately 7 million bbl per day. A methanol coal plant with a production of 120,000 bbl per day would cost $1.8 billion. The construction of an alcohol industry capable of supplying the motor fuel requirements of a large percentage of the U.S. vehicle fleet would be extremely expensive.

The Brazilian Experience

ETHANOL PRODUCTION Brazil's automobile fleet increased from 235,000 to over 8 million vehicles in the past thirty years. Brazil, with a yearly production in excess of one million vehicles, is the ninth largest car manufacturer in the world. Unfortunately, this expansion in the automotive industry has not been matched by a parallel growth in the oil industry. Domestic oil production met 80 percent of the demand in 1950, but at the present time only 17 percent of Brazil's oil requirements are supplied by domestic crude oil. Dependence on imported crude oil and the rapid increase in oil prices have seriously affected Brazil's balance of payments.

Brazil has now organized a program to develop an ethanol motor fuel industry. Initial target production was 800 million gal of ethanol per year by 1980. Current production plans are 2.8 billion gal of ethanol per year. Most of the ethanol plants are connected with sugar distilleries and use sugarcane as feedstock. The average plant production is 3,200 gal per day, or 1 million gal per 300-day year. Each plant requires an initial investment of $6 million and a supporting investment of $3 million in the agricultural sector.

ETHANOL AUTOMOTIVE CONVERSIONS The Brazilian government has signed agreements with the automotive industry for the production of full alcohol-powered cars and the conversion of existing vehicles to ethanol. In the period 1980–1982, 900,000 new ethanol powered vehicles will be produced, and 170,000 used cars will be converted to ethanol. Government incentives are available to private industry for implementation of the alcohol program.

Advantages of Alcohols

1. Alcohols could replace fuels derived from crude oil and allow some nations without crude oil reserves to become self-sufficient in energy.
2. The use of ethanol could increase and stabilize farm income.
3. Alcohols have a pump octane rating of 99–102 compared with unleaded gasoline, which has an octane rating of 90.
4. Alcohols act as an octane booster when they are mixed with gasoline. A mix of 10 percent methanol and gasoline would have an octane rating of 95.
5. Alcohols with higher octane ratings would permit the use of engines with higher compression ratios. With higher compression ratios, more power can be obtained from an engine. In other words, the equivalent power output would be obtained from smaller engines powered with alcohols.
6. Alcohols, with their greater latent heat of vaporization, result in more compression work from an engine. The increased cooling of the air-

fuel mixture allows more mixture to be packed into the cylinders. The use of pure methanol could increase the power output of an engine by 10 percent compared with the same engine fueled with gasoline. Methanol has an ideal air-fuel ratio of 6.4:1, whereas the ideal gasoline air-fuel ratio is 14.5:1. Although alcohols allow greater compression work from an engine than gasoline, they have lower BTUs per gallon than gasoline. The net result is very little difference in fuel economy between alcohol, gasoline mixes, and straight gasoline. (Fuel economy with alcohols is discussed in the next chapter.)
7. Alcohols have greater leaning capabilities than gasoline.

Disadvantages of Alcohols

1. Because alcohols act as cleaning agents, filters can become plugged in gasoline fuel systems converted to alcohol.
2. Alcohols and gasoline separate if they are contaminated with small amounts of water, especially in cold weather. Phase separation of alcohol and gasoline occurs more easily with methanol.
3. Some gasoline fuel system parts are not compatible with alcohols, especially the terne plating in the fuel tank. Small percentages of ethanol mixed with gasoline will experience fewer compatibility problems, whereas pure methanol would cause the most serious problems, including damage to parts of the fuel system.
4. Because alcohols do not vaporize as easily as gasoline at low temperatures, hard-starting problems can occur with the use of alcohols in cooler climates. Cold-starting problems with pure methanol could occur at temperatures below 50°F (10°C). Some cures for cold-starting problems encountered with alcohols are:
   a. Blending volatile components with alcohol.
   b. Using an auxiliary starting fuel.
   c. Using electric fuel vaporizers.
   d. Improving fuel vaporization with a fuel injection system rather than a carburetor.
5. Alcohols, with their greater latent heat of vaporization, may require more intake manifold heat to prevent drivability problems.
6. Alcohols will absorb moisture from the atmosphere. Special venting on fuel tanks and carburetor float bowls will be required with alcohol fuels.

7. Since alcohols would require a whole new production and distribution system, billions of dollars would have to be spent if alcohols were to replace gasoline.
8. Alcohols require special care in transportation and distribution because they tend to absorb moisture.

Questions

1. Ethanol and methanol contain oxygen.   T   F
2. The BTU content of ethanol is _____ per gallon.
3. The most common U.S. feedstock for ethanol production is _____.
4. Gasohol is a mixture of _____ percent ethanol and gasoline.
5. When ethanol is produced from corn, yeast is used to convert the sugar to alcohol.   T   F
6. Alcohol rated at 195 proof would contain 10 percent water.   T   F
7. Gasohol may be produced with 190 proof ethanol.   T   F
8. Methanol has approximately _____ times the cooling effect of gasoline when it changes from a liquid to a vapor.
9. Methanol can be produced by coal gasification.   T   F
10. The same fuel tank could be used if a gasoline fuel system was converted to pure methanol.   T   F
11. Alcohol fuels could cause hard-starting problems at 0°F (−18°C).   T   F

CHAPTER 13

Engine Tuning, Fuel Economy, and Exhaust Emissions with Alcohol Fuels

Fuel System And Ignition Tuning

AIR-FUEL RATIO REQUIREMENTS Methanol has an ideal air-fuel ratio of 6.4:1. The ideal air-fuel ratio for ethanol is 9:1. Compared with gasoline, which has an ideal air fuel ratio of 14.5:1, alcohol air-fuel ratios are much richer. A mixture of 10 percent ethanol and gasoline should perform satisfactorily without altering the carburetor. As the percentage of alcohol is increased in the alcohol-gasoline mixture, the air-fuel ratio delivered by the carburetor must be enriched.

CARBURETOR MODIFICATIONS Variable main metering jets as pictured in Figure 13-1 are available to provide the necessary mixture enrichment for alcohol fuels. In computer-controlled carburetor systems, as described in Chapter 6, the oxygen sensor and computer will automatically correct the air-fuel ratio for alcohol-gasoline mixes containing low percentages of alcohol. Computer-controlled carburetors or injection systems could be engineered to automatically provide the correct air-fuel ratio with much higher percentages of alcohol mixed with gasoline.

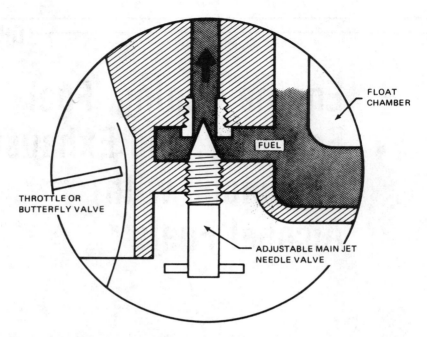

FIGURE 13-1. Variable Main Metering Jets. *(Courtesy of Desert Publications)*

Problems with carburetor jet sizes can occur in countries like Brazil where different mixtures of ethanol and gasoline may be available. A blend of 50 percent ethanol requires an air-fuel ratio of approximately 12:1. Fuel economy decreases by about 14 percent with a 50 percent blend of ethanol and gasoline and carburetor jets that provide a 12:1 air-fuel ratio. When the same vehicle is operated on a 20 percent ethanol and gasoline mix, fuel economy decreases by 20 percent, and HC and CO levels increase. The electronic lean limit system in Figure 13-2 was developed for use in Brazil to automatically provide the correct air-fuel ratio for ethanol-gasoline blends containing up to 50 percent ethanol.

A magnetic sensor positioned near the flywheel ring gear teeth senses flywheel acceleration. The rate of change in flywheel acceleration is proportional to the quality of combustion. Each combustion event is evaluated as "good" or "bad" by the magnetic sensor and control unit. The carburetor jet size is designed to provide the rich mixture required by the highest percentage of ethanol that would be used in the fuel. A specific amount of air flows past the servo bleed port into the intake manifold. The position of the servo bleed screw supplies the necessary air flow to provide the correct air-fuel ratio for different ethanol-gasoline mixes. The control unit operates two solenoid valves that supply a specific amount of vacuum to the servo. If a lean, "bad" combustion signal is received by the magnetic sensor, the control unit and solenoid valves will decrease the vacuum supplied to the servo. The servo bleed port will close, less air will enter the bleed passage, and the mixture will become richer. The vacuum regulator prevents pulsations in the manifold vacuum from affecting the system. Fuel economy remains more constant in relation to the amount of ethanol used in the fuel with the lean limit system.

The idle mixture screws would require adjustment when changing to higher percentages of alcohol fuels. As mentioned in the previous chapter, additional intake manifold heating may be required with alcohol fuels, especially in cold weather. Some carburetor and fuel pump components may not be compatible with alcohol-gasoline mixes containing high percentages of alcohol.

IGNITION TUNING Ignition timing can be advanced when converting from gasoline to alcohol fuels because of the higher octane rating of the alcohols. The amount of additional timing advance depends on the percentage of alcohol in the gasoline-alcohol mix.

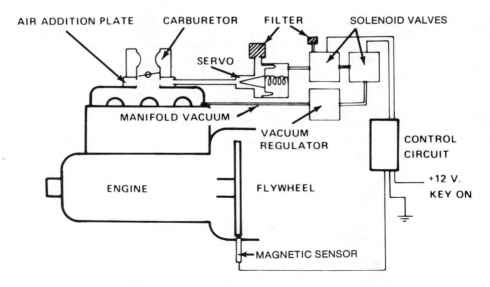

FIGURE 13-2. Lean Limit Control System. *(Reprinted with permission © 1980, Society of Automotive Engineers)*

ALCOHOLS IN DIESEL ENGINES Diesel fuel has a cetane rating of 45. Cetane rating indicates the self-ignition qualities of a fuel. Alcohols are not very compatible with diesel engines because they have a cetane rating of zero. When alcohols are mixed with diesel fuel, it is difficult to prevent separation of the alcohol from the diesel fuel. Other liquids can be mixed with diesel fuel and alcohol to form an emulsion that prevents separation. Some manufacturers have developed a separate injection system for the alcohol, but the double injection system is very expensive. Alcohols can be injected into the air intake on the diesel engine.

One of the most recent developments involves addition of alcohol to the injection lines of a diesel engine. A one-way check valve is used to allow alcohol into the injection line and prevent any reverse flow of diesel fuel into the alcohol system. A low-pressure pump forces ethanol from a separate fuel tank to the one-way check valves, as illustrated in Figure 13-3.

Negative pressure occurs in the injection line the instant the injector closes. The negative pressure will open the one-way check valve and

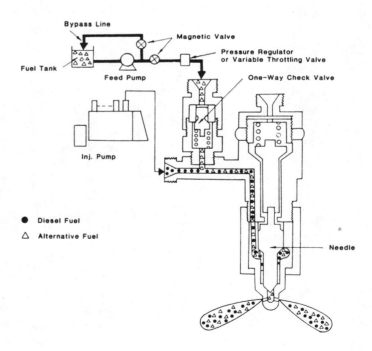

FIGURE 13-3. Alcohol Diesel Fuel System. *(Reprinted with permission © 1981, Society of Automotive Engineers)*

move alcohol into the injection line. The next time the injector is opened by diesel injection pump pressure, diesel fuel and alcohol will be injected into the combustion chamber.

Fuel Mileage And Emission Levels With Alcohols

FUEL ECONOMY The fuel mileage and exhaust emission test results on the following pages are taken from a study of alcohol motor fuels conducted at the University of Santa Clara, California. A fuel-injected Toyota Supra was used in the tests. Figure 13-4 indicates the fuel mileage obtained with gasoline and different blends of alcohol and gasoline. Test results indicate very little difference in fuel mileage with gasoline or gasoline and 10 percent alcohol. Fuel mileage decreases as higher percentages of alcohol are mixed with gasoline because air-fuel ratios must be richer.

EVAPORATIVE EMISSIONS Evaporative emissions from the fuel tank and carburetor show a slight increase when higher percentages of alcohols are mixed with gasoline, as pictured in Figure 13-5.

NITROUS OXIDE EMISSIONS With a mixture of 30 percent ethanol or methanol and gasoline, nitrous oxide (NOx) emissions are more than double the NOx emissions with gasoline. The maximum U.S. federal NOx

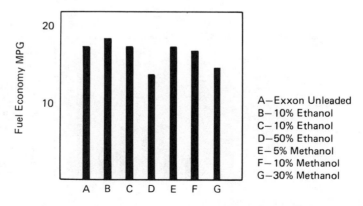

FIGURE 13-4. Alcohol Fuel Economy. *(Reprinted with permission © 1980, Society of Automotive Engineers)*

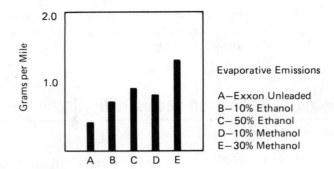

FIGURE 13-5. Alcohol Evaporative Emissions. *(Reprinted with permission © 1980, Society of Automotive Engineers)*

emission level is 1 gram per mile (gpm). As indicated in Figure 13-6, the use of over 30 percent alcohol mixed with gasoline would require some additional emission control equipment to lower NOx emissions.

CARBON MONOXIDE EMISSIONS Carbon monoxide emission levels improve when alcohol-gasoline mixes are used in place of gasoline, as shown in Figure 13-7. The U.S. federal requirement for CO levels has been revised from 3.4 gpm to 7 gpm.

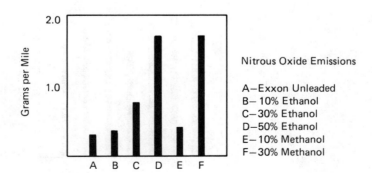

FIGURE 13-6. Alcohol Nitrous Oxide Emissions. *(Reprinted with permission © 1980, Society of Automotive Engineers)*

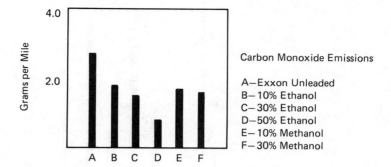

FIGURE 13-7. Alcohol Carbon Monoxide Levels. *(Reprinted with permission © 1980, Society of Automotive Engineers)*

HYDROCARBON EMISSIONS When alcohol-gasoline mixes are substituted for gasoline, hydrocarbon emission levels increase slightly.

With 30 percent methanol or 50 percent ethanol, HC emission levels are below the U.S. federal standard of 0.41 gpm, as illustrated in Figure 13-8.

ALDEHYDE EMISSIONS When higher percentages of alcohols are mixed with gasoline, aldehyde emissions increase proportionately, as indicated in Figure 13-9.

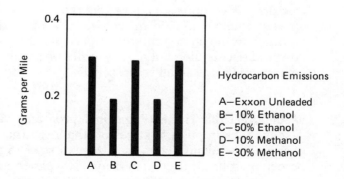

FIGURE 13-8. Alcohol Hydrocarbon Emission Levels. *(Reprinted with permission © 1980, Society of Automotive Engineers)*

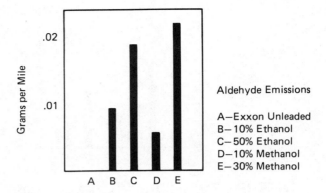

FIGURE 13-9. Alcohol Aldehyde Emission Levels. *(Reprinted with permission © 1980, Society of Automotive Engineers)*

Savannah River Plant Alcohol Program

FLEET SIZE The Savannah River Plant (SRP) is operated by the Du Pont Company for the U.S. Department of Energy to produce nuclear materials for national defense and peacetime projects. Automotive units operated by the SRP include 300 sedans and station wagons, 500 light trucks, and 100 medium and heavy trucks. The fleet also contains 250 nonautomotive units such as welders, pumps, forklifts, mowers, farm tractors, air compressors, and generators. Automotive fleet mileage is approximately 9 million mi per year. Total fuel consumption by the automotive and nonautomotive fleets is 900,000 gal per year. All gasoline-powered fleet equipment now operates on gasohol, except for two cycle engines and some small equipment.

GASOHOL BLENDS The SRP gasohol program was initiated to reduce gasoline consumption. Mixtures of ethanol and gasoline containing 10–20 percent ethanol have been used. A mixture of 15 percent ethanol and 85 percent gasoline is used at present. Fleet experience includes 1.8 million mi on 180,000 gal of 10/90 gasohol (10 percent ethanol and 90 percent gasoline), 5.8 million mi on 580,000 gal of 15/85 gasohol, and 2.4 million mi on 240,000 gal of 20/80 gasohol.

PROBLEMS ENCOUNTERED

1. Problems were encountered with incomplete mixing of the alcohol and gasoline when the alcohol was dumped into the gasoline in underground storage tanks. These problems were solved by always dumping the gasoline into the alcohol.
2. Plugged in-line fuel filters were the most frequent problem, which usually occurred in the first tank of gasohol. Several older vehicles required two or three filter replacements, and gas tank removal and purging were necessary on two older vehicles.
3. Carburetor accelerator pumps had to be replaced on many Dodge pickup trucks. Rubber pump plunger cups were not compatible with gasohol. Fewer problems occurred with neoprene plunger cups.
4. Flexible rubber fuel lines tended to deteriorate.
5. Timing had to be advanced on some vehicles to obtain satisfactory performance.
6. Carburetor jets on several vehicles became plugged with resin particles. Electrical components are protected by a resin coating in the fuel pump chamber. Ethanol deteriorated the resin coating and small particles were carried through the fuel system into the carburetor. The electric fuel pumps were replaced with diaphragm pumps.
7. Approximately twenty-five vehicles experienced problems with phenolic resin carburetor floats, which seemed to absorb gasohol more easily than gasoline. Carburetor flooding occurred when the float became saturated with gasohol. Replacement floats were made from the same material. Tests have shown that floats coated with Du Pont "Viton" absorb less gasohol than uncoated floats. The carburetor float problem was most noticeable with 20/80 gasohol.

COST FACTORS The cost of ethanol to the SRP was $1.87 per gal, and the cost of gasoline was $1.07 per gal. A 15/85 gasohol blend cost $1.19 per gallon, and the additional cost for 15/85 gasohol compared with gasoline was 12¢ per gallon.

CONCLUSIONS

1. No significant change in performance of automotive or nonautomotive equipment was experienced.
2. Fuel economy was the same with gasohol or gasoline.
3. Maintenance problems were fewer than expected and relatively inexpensive with gasohol.

4. The cost of gasohol is significantly higher than gasoline costs, but the cost difference should decline if gasoline costs increase.
5. Gasohol utilization can reduce gasoline consumption.

Engine Wear and Methanol Fuel

U.S. ARMY TEST RESULTS The U.S. Army Fuels and Lubricants Research Laboratory in San Antonio, Texas, has done a considerable amount of research on engine wear with methanol fuel. In a 462-hour test with methanol fuel, top ring gap increase was 0.006 (0.015 cm) and second ring gap increase was 0.008 (0.020 cm). Cylinder wear at the top ring travel was 0.0004 (0.0010 cm). An engine fueled with gasoline for the same time period showed a top ring gap increase of 0.002 (0.005 cm), and a second ring gap increase of 0.003 (0.008 cm). The gasoline-fueled engine did not experience any cylinder wear. Additional ring and cylinder wear with methanol fuel was attributed to corrosive action and oil degradation by the methanol.

Questions

1. The ideal air-fuel ratio with methanol fuel is _____.
2. Variable main metering jets may be used to provide the correct air-fuel ratio with alcohol fuels.   T   F
3. Alcohols have good self-ignition qualities.   T   F
4. Fuel economy decreases as the percentage of ethanol is increased in a gasoline-ethanol mix.   T   F
5. Emissions of nitrous oxides are higher with a 30 percent ethanol and gasoline mix compared with straight gasoline fuel.   T   F
6. Carbon monoxide emission levels increase in proportion to the amount of ethanol mixed with gasoline.   T   F
7. Phenolic resin carburetor floats may become _____ with gasohol fuel.
8. Engine wear will be the same with gasoline or methanol fuel.   T   F

CHAPTER 14

Hydrogen—The Fuel of the Twenty-First Century

Hydrogen Development

GENERAL FACTS Hydrogen is one of the most abundant elements in the world. Hydrogen combines with other elements to form compounds. Water is a compound composed of hydrogen and oxygen. Hydrogen weighs 0.6 lb (0.270 kg) per U.S. gal and contains 30,000 BTUs per gal. Each pound of hydrogen contains 51,600 BTUs. Liquid hydrogen must be cooled below the boiling point of $-423°F$ ($-252°C$). The chemical symbol for hydrogen is H_2. Liquid hydrogen and gaseous hydrogen are identified as LH_2 and GH_2. Because hydrogen is so abundant, many experts in the energy field believe the era of hydrogen fuel is inevitable.

SOURCES OF HYDROGEN Steam can be used to reform natural gas (CH_4) to hydrogen (H_2). This source of hydrogen is a nonrenewable natural resource. It would be more economically feasible to use natural gas itself as a fuel. Hydrogen can also be produced through the gasification of coal. Although there are huge reserves of coal in the United States, the gasification of coal creates pollution and consumes a nonrenewable natural resource. Still another method of producing hydrogen is to use an electric current to separate the hydrogen and oxygen in water. This process is referred to as electrolysis of water. Using nuclear-generated electricity to produce hydrogen by electrolysis would appear to involve fewer trade-offs between cost, pollution, and natural resource consumption.

COST FACTORS The costs of hydrogen production were detailed earlier in Figure 1-3. Producing hydrogen from advanced special electrolysis (nuclear-generated electricity) costs $16 per million BTUs and involves a 27 percent efficient energy conversion. Liquefaction of hydrogen requires a considerable amount of process energy and costs $5-8 per million BTUs. Crude oil prices would have to be $50-60 (U.S.) per bbl for hydrogen to compete in price with gasoline.

HYDROGEN FROM SOLAR ENERGY Tokio Ota, professor of engineering at Yokohama National University, has been experimenting with hydrogen production from solar energy. Large ocean rafts are mounted with salt-resistant, plastic parabolic mirrors to collect the sun's heat. The solar heat is used to boil water, and the resulting steam drives turbines to generate electricity. Seawater is separated into hydrogen and oxygen by electrolysis. The gases are supercooled to create liquid hydrogen and oxygen.

HYDROGEN AS AN AIRCRAFT FUEL Lockheed Aircraft Corporation has been working on the development of liquid hydrogen as an aviation fuel. Weight reduction due to the much lighter hydrogen fuel would be the most important advantage of a hydrogen-powered aircraft. Daniel Brewer, a Lockheed engineer, has estimated that a hydrogen-powered aircraft with a 25-ton payload would weigh 144 tons. A conventional jet aircraft with the same payload weighs 195 tons. Smaller, lighter engines exploiting hydrogen's cooling properties could reduce fuel consumption by 5-10 percent. Brewer also estimates that a 25 percent reduction in engine maintenance and a 25 percent increase in engine life would be obtained with hydrogen fuel. With these potential savings, airline operators could find the conversion to hydrogen fuel economically feasible long before the costs of conventional jet fuel and LH_2 reached parity on a BTU basis. Crucial problems to be overcome in the use of LH_2 as an aircraft fuel include metal embrittlement by LH_2, the super insulation of fuel tanks, and the production of more durable LH_2 pumps.

Hydrogen As An Automotive Fuel

FUEL STORAGE With present technology, the two most feasible methods of storing hydrogen on a vehicle are hydride storage or liquid storage.

Hydride storage involves filling two-thirds of a tank with iron titanium, a crushed pebblelike substance, and then pumping hydrogen gas into the tank. The iron titanium absorbs the hydrogen just as a sponge soaks up water. Iron titanium hydride is formed when the hydrogen is absorbed by iron titanium. When heat from the engine cooling or exhaust system is applied to the tank, hydrogen is released from the iron titanium. Hydride tanks filled with hydrogen have an excellent safety level. Even bullet punctures will not cause hydride tanks to explode. The disadvantage of hydride tanks is their excessive weight. A hydride tank with a 200-mile driving range for a small car weighs 600 lb (270 kg). Engineers from many countries involved in hydrogen projects are searching for a lighter hydride capable of storing hydrogen at greater densities. Misch-metal hydrides, capable of storing hydrogen at volume densities greater than liquid hydrogen, have recently been developed.

Liquid hydrogen storage involves the use of cryogenic technology very similar to the storage requirements for liquid natural gas described in Chapter 11. Liquid hydrogen tanks have double walls and special insulation between the walls. The insulated space is maintained in a permanent vacuum.

LIQUID HYDROGEN PROJECT Hydrogen is used in the gasoline refining process. Thus far, hydrogen fuel has been widely used only in space rockets. Hydrogen as an automotive fuel is in the experimental stage. A recent test on a hydrogen-powered vehicle was conducted at the Los Alamos Scientific Laboratory in Los Alamos, New Mexico. The vehicle used in the experiment was a 1979 Buick Century equipped with a 231 CID V6 turbocharged engine. Modifications to the stock vehicle are listed below.

1. The gasoline tank and some sheet metal components of the trunk were removed to make room for the LH_2 tank.
2. All emission-control equipment, including the catalytic converter, was removed.
3. The trunk area was sealed from the passenger compartment.
4. Ventilation of the trunk was accomplished by installing louvres in the trunk lid.
5. Hydrogen detecting devices were installed in the trunk and passenger compartment.
6. An Impco 300 propane mixer was installed on top of the gasoline carburetor.
7. Water was pumped from a trunk-mounted water tank to the gasoline

carburetor by the conventional gasoline fuel pump. Water induction through the gasoline carburetor prevents manifold backfiring. Carburetor idle jets were plugged to prevent water induction during idle operation. Water induction also lowers nitrous oxide (NOx) emission levels.

8. A complete hydrogen fuel system was installed, including a pressure regulator mounted under the hood and all necessary connecting hoses, as shown in Figure 14-1.
9. Conventional plug wires were replaced with solid wires.
10. Vacuum and centrifugal advances were removed and ignition timing was set at a fixed 20° BTDC.
11. An oxygen sensor was installed in the exhaust system to indicate the air-fuel ratio to the operator.
12. Digital fuel level and fuel tank pressure indicators were installed.
13. A thermostat rated at 158°F (70°C) was installed in place of the original thermostat.
14. Additional gauges installed included a tachometer, manifold pressure gauge, and a hydrogen pressure gauge to monitor pressure to the mixer.

FIGURE 14-1. Hydrogen Fuel System. *(Reprinted with permission © 1981, Society of Automotive Engineers)*

FIGURE 14-2. Liquid Hydrogen Fuel Tank. *(Reprinted with permission © 1981, Society of Automotive Engineers)*

The onboard LH_2 fuel tank illustrated in Figure 14-2 has a maximum working pressure of 65 psi. Fuel tank capacity is 29.1 gal of LH_2, which has the energy equivalent to 7.8 gal of gasoline. Hydrogen boil-off rate at the lockup pressure is 3.5 percent per day. Fuel tank driving range is about 170 mi (272 km).

Vented hydrogen gas is released through a small hose at the rear of the vehicle. A catalytic combustion unit for the vent hose is being evaluated. The catalytic unit would oxidize hydrogen gas to water vapor before being released from the vent hose. At higher tank pressures, gaseous hydrogen is delivered to the fuel system. Liquid hydrogen is utilized at lower tank pressures. A hydrogen heat exchanger in the fuel tank brings GH_2 or LH_2 up to the ambient temperature before delivering the hydrogen into the fuel system. Engine coolant is circulated through the heat exchanger. The LH_2 and GH_2 modes of operation are very similar to the vapor and liquid modes of operation in the liquid natural gas system described in Chapter 11. Fuel tank weight is 184.4 lb (83.6 kg).

The test car and refueling station are illustrated in Figure 14-3. Refueling steps are listed in the following sequence.

FIGURE 14-3. Liquid Hydrogen Refueling Station. *(Reprinted with permission © 1981, Society of Automotive Engineers)*

1. Fill and vent return connections are completed between the filling station and the vehicle tank, and the start button is pressed.
2. After an automatic line evacuation operation and "connections tight" check, a readiness indication is displayed and the two manual tank valves are opened.
3. Tank refueling continues until the completion of filling signal is received.
4. Manual tank valves are closed and lines are evacuated and nitrogen purged. Lines are manually disconnected.

If an unsafe condition exists during the refueling operation, the sequence will be stopped automatically and the problem signaled to the operator. Warm tank refueling time is 40 minutes, and 9 minutes are required to fill a cold tank. Fuel mileage was 5.7 mpg (2.4 km/l) on liquid hydrogen, which is equivalent to 21 mpg (8.9 km/l) on gasoline.

Project Summary

1. The vehicle was operated for 1,700 miles (2,700 km) on hydrogen fuel without any safety problems.

2. Fuel tank heat loss, capacity, and weight could be improved.
3. Power loss of 20 percent makes acceleration performance disappointing. Occasional intake manifold backfiring occurred when the engine was started or was operated under heavy load conditions. Conversion to direct hydrogen injection could solve these problems.
4. Some turbocharger damage was experienced when water droplets, from the water injection, hit the compressor blades.

HYDROGEN INJECTION PROJECT The Musashi Institute of Technology in Tokyo, Japan, has succeeded in developing hydrogen fuel injection. One of their latest projects has been to design a hydrogen fuel injection system on a two-cycle, three-cylinder 1.1 liter engine. The basic hydrogen fuel system is shown in Figure 14-4.

The LH_2 fuel tank is similar to the cryogenic tank described previously in this chapter. Tank capacity is 82 l and empty tank weight is 77 lb (35 kg). An electric LH_2 pump is mounted in the fuel tank. Pump pressure forces liquid hydrogen from the tank through the heat exchanger to the GH_2 chamber. Liquid hydrogen changes to a vapor in the heat exchanger. Pressure pulsations of the GH_2 are smoothed out by the insulated 1-l GH_2 chamber. Engine coolant is circulated through the heat exchanger. Heat-insulated pipes conduct the gaseous hydrogen from the GH_2 chamber to the injectors.

The hydrogen injectors are illustrated in Figure 14-5. The camshaft moves the rocker arm vertically in order to open the injector. Engine speed is controlled by the amount of fuel injected.

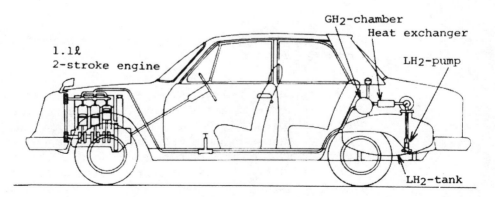

FIGURE 14-4. Hydrogen Fuel System. *(Reprinted with permission © 1982, Society of Automotive Engineers)*

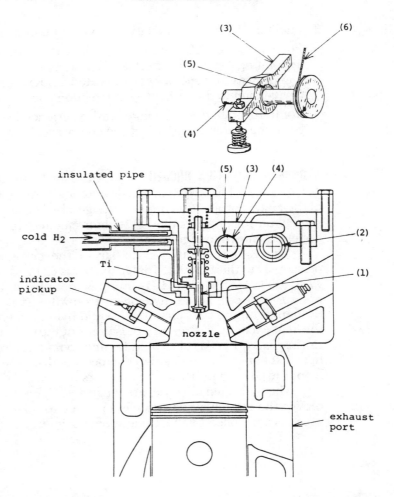

FIGURE 14-5. Hydrogen Injectors. *(Reprinted with permission © 1982, Society of Automotive Engineers)*

The amount of injector opening is determined by an eccentric, item 5, between the rocker arm shaft and the rocker arm. Rotation of the rocker arm shaft and eccentric is accomplished by a pulley and cable, item 6. The cable is connected to the accelerator pedal. Movement of the accelerator pedal rotates the rocker arm shaft and eccentric bushing fastened to the shaft. In this way, accelerator pedal movement controls the injector opening and engine speed. The two-cycle engine uses intake and exhaust ports in the block. Gaseous hydrogen at −30°C to −50°C (−34 to −45°F) is injected on the compression stroke. The engine compression ratio is 9:1.

Project Results

1. Engine performance improved 20–30 percent compared with the same engine fueled with gasoline. It was discovered that engine performance was greatly influenced by the direction of injected fuel. A rain shower injector spray pattern with uniform distribution was found to provide best performance. Performance was improved when the spray pattern did not collide against the combustion chamber or cylinder walls.
2. Backfiring was eliminated completely because the cold fuel was injected on the compression stroke.
3. Low nitrous oxide (NOx) emissions were also attributed to the cold injected fuel.
4. The project did not measure fuel economy.

SLX HYDROGEN BREAKTHROUGH The hydrogen-fueled vehicles discussed in the preceding sections require expensive hydrogen filling facilities. With the high cost of liquefying hydrogen, the ideal system would allow the owner to fill his fuel tank with water. An onboard separator would disassociate the hydrogen from the oxygen in the water, and the engine would be fueled with hydrogen. This has long been the dream of many car owners, engineers, and scientists! The SLX hydrogen fuel system developed by Omnia Research Corp. in California may be the long-awaited answer to the dream.

Water, H_2O, contains hydrogen and oxygen atoms. Two hydrogen atoms and one oxygen atom are combined by a form of magnetic bonding in each molecule of water, as indicated in Figure 14-6.

If the water molecules can be placed in a situation where the oxygen atoms will be attracted away from the hydrogen atoms, the hydrogen atoms will be left by themselves. If water is properly conditioned in terms of temperature and pressure and subjected to a particular chemical reactant, the oxygen atoms can be attracted away from the hydrogen atoms. In the SLX system, water is heated to 300°F (149°C). Steam is introduced to a reactant chamber where the reactant captures and holds oxygen atoms while releasing hydrogen atoms. Specific temperatures and pressures must be maintained in the reactant chambers. The hydrogen atoms are reunited by a photochemical process to form H_2 in the reactant chamber after they are freed from the oxygen atoms. Reaction heat is generated when the oxygen atoms are separated, and reaction heat is released when the photochemical process reunites the hydrogen atoms to form H_2. The SLX hydrogen system with reactant chambers A, B, and C is illustrated in Figure 14-7.

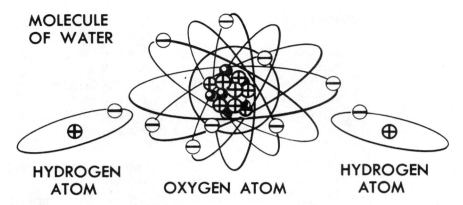

FIGURE 14-6. Water Molecule. *(Courtesy of General Motors of Canada, Ltd.)*

Reactant chamber cycles of 6 seconds duration are controlled by cam-controlled diverter valves. A rotating camshaft operates the diverter valves at the right instant. Three cycles are repeated alternately in each reactant chamber. The first cycle consists of the introduction of water and the production of hydrogen. Free oxygen and surplus hydrogen are purged during the second cycle. During the third cycle, a vacuum is introduced to the reactant chamber in preparation for the first cycle. A Plymouth Horizon TC3 with an SLX hydrogen system is pictured in Figures 14-8 and 14-9. No fuel economy or performance figures from the project are available. Information on the underhood components in the system has not been released yet.

EXHAUST EMISSIONS WITH HYDROGEN FUEL The main tailpipe emission with hydrogen fuel is water vapor. With hydrogen fuel, emissions of CO and HC are nonexistent. Nitrous oxide (NOx) emissions can be controlled with correct air-fuel ratios.

SAFETY AND HYDROGEN FUEL The safety aspects of any fuel are a major concern. The excellent safety features of hydride tanks have been mentioned already. Gasoline or diesel fuel will puddle or spread when a leak occurs, but hydrogen escaping from a leak would go straight up and disperse in the atmosphere because the hydrogen atoms are very light. As well, the safety records of hydrogen fuel might be better than those of hydrocarbon fuels.

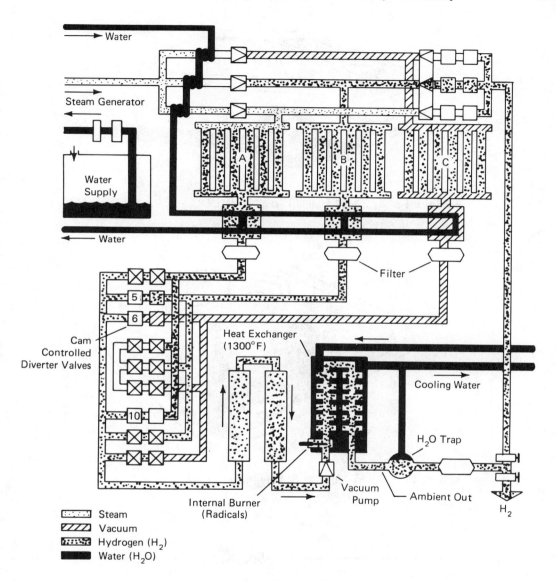

FIGURE 14-7. SLX Hydrogen System.

Alternate Automotive Fuels

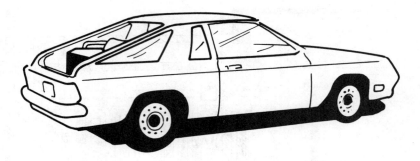

FIGURE 14-8. Hydrogen-Fueled Plymouth Horizon.

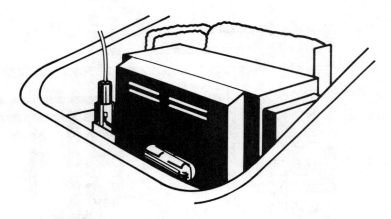

FIGURE 14-9. Trunk-Mounted SLX Hydrogen System.

―――――――――――――― Questions ――――――――――――――

1. Hydrogen fuel is lighter than hydrocarbon fuels.   T   F
2. The BTU content of hydrogen is _____ BTUs per U.S. gallon.
3. Liquid hydrogen must be maintained at _____ degrees F.
4. The most efficient method of hydrogen production is by electrolysis of water utilizing electricity from a coal-fired generating station.   T   F

5. One of the most important advantages of hydrogen as an aircraft fuel would be _____ _____.
6. Hydride hydrogen fuel tanks are suitable for small vehicles.   T F
7. One of the most common problems with hydrogen fuel systems has been intake manifold _____.
8. Using a hydrogen injection system on a two-cycle engine can eliminate intake manifold backfiring.   T F
9. Hydrogen refueling facilities are eliminated with the SLX hydrogen system.   T F

CHAPTER 15

Future Energy Sources

Solar Energy

POTENTIAL OF SOLAR POWER More potential energy exists in solar power than in any other source of energy. If we could collect and use 1/10,000 of the solar energy available in the world, all our energy needs would be met. Without the sun's warming effect, the temperature of our planet would never rise above −450°F (−266°C). Each year the sun drenches the United States with 500 times the total energy that we consume. If the sun shone on only 2 percent of the nation's surface and we could tap one-tenth of the energy available, the nation's energy needs would be met. Solar energy has a multitude of applications. The development of the Stirling engine could produce solar-generated electricity in the future. Parabolic dish concentrators apply solar heat to a working fluid, as illustrated in Figure 15-1.

Solar heat expands the working fluid and drives the pistons in the Stirling engine. The engine is used to drive an electric generator. A free piston Stirling engine is illustrated in Figure 15-2.

Many technological breakthroughs will occur in the next decade to utilize solar energy in various fields. The Stirling engine is being studied as a possible power plant in the automotive industry. Figure 15-3 shows the main components in a solar thermal power system.

Wind Energy

DEVELOPMENT The U.S. Wind Energy Systems Act of 1980 has initiated an eight-year, $900 million program to develop cost-effective wind power

FIGURE 15-1. Parabolic Solar Concentrators. *(Reprinted with permission © 1981, Society of Automotive Engineers)*

systems. A number of large wind turbines are already in the experimental and developmental stages. It would require 30,000 large turbines and thousands of smaller ones to supply 10 percent of the U.S. electrical power requirements by the year 2000. Electric-powered vehicles may be introduced by the automotive industry in the near future. Wind-generated electricity may be charging the batteries in our future electric vehicles. Figure 15-4 pictures a wind generator.

Technology From The Sun

THE FUSION SOLUTION Virtually unlimited energy is available by fusion. The development of fusion energy represents one of the greatest

Future Energy Sources

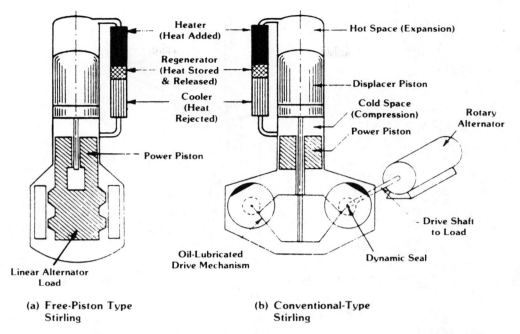

FIGURE 15-2. Free Piston Stirling Engine. *(Reprinted with permission © 1981, Society of Automotive Engineers)*

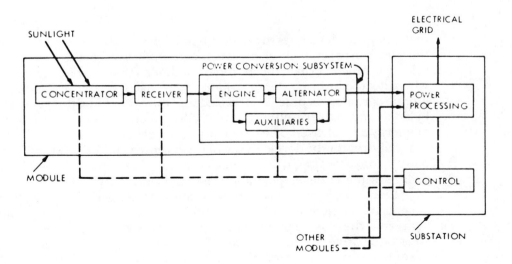

FIGURE 15-3. Solar Thermal Power System. *(Reprinted with permission © 1981, Society of Automotive Engineers)*

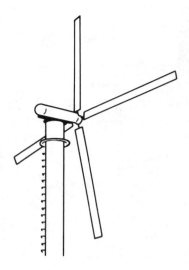

FIGURE 15-4. Wind Generator.

technological challenges to face mankind. Fusion is the process by which the sun burns and gives off heat continuously. To achieve fusion, scientists must recreate the extreme conditions that exist inside the sun, where heat reduces matter to an ionized gas called plasma. In this plasma, atomic particles become so energized that they overcome their electrical balance, and in the resulting atomic collisions and fusing, large amounts of energy are released. The fusion process is self-sustaining. U.S. budget expenditures for the development of fusion energy totaled $800 million in one recent year. The Tokamak device pictured in Figure 15-5 is a fusion generator under development at Princeton University.

Electromagnetic coils, item A, create a magnetic field to contain the plasma, component B. Helium and lithium layers, items C and D, carry heat from the plasma to a heat exchanger. The heat exchanger creates steam to drive electric turbines. Fusion energy has tremendous advantages over present fission reactors. Only a fraction of the radioactivity in fission generators is present in fusion generators. Fusion technology is extremely complex and expensive. The commercial fusion solution will not come before the year 2000.

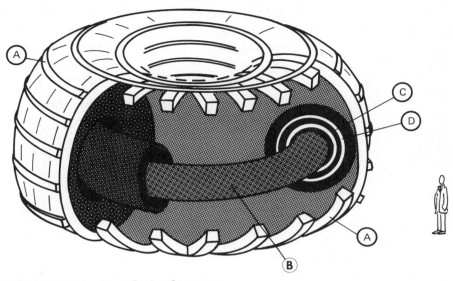

FIGURE 15-5. Tokamak Fusion Generator.

Energy From The Sea

OCEAN THERMAL ENERGY CONVERSION The ocean absorbs about 75 percent of the solar energy that strikes the earth. Ocean thermal energy conversion (OTEC) is a method of tapping this energy by making use of the temperature difference between sun-warmed surface waters and cold ocean depths. Warm surface water vaporizes a fluid with a low boiling point such as ammonia. The vapor drives a turbine to generate electricity. Cold water pumped from a level of 3,000 ft (912 m) condenses the vapor back into liquid. OTEC is pollution-free, does not require any land space, and does not consume any hydrocarbon fuels. The U.S. Department of Energy estimates that OTEC will replace 400,000 bbl of oil per day by the year 2000.

―――――――――― Questions ――――――――――

1. Without the sun's warming effect, the temperature of the earth would be about _____ degrees F.

2. All the world's energy needs could be met if we could collect and utilize _____ of the available solar energy.
3. Solar energy can be used to operate the Stirling engine. T F
4. Fusion-powered generators have an improved safety factor compared to present nuclear fission generating plants. T F
5. The United States has spent $_____ per year on the development of fusion energy.
6. List three advantages of ocean thermal energy conversion.
   1. _____
   2. _____
   3. _____

Index

A

Air fuel ratios
   alcohols, 199-201
   compressed natural gas, 146, 147
   propane, 109, 110
Alcohols
   advantages, 196, 197
   air fuel ratios, 199
   Brazilian experience, 195, 196
   carburetor modifications, 199-201
   diesel-alcohol systems, 202
   disadvantages, 197, 198
   emission levels, 203-206
   ethanol, 189-194
      chemical composition, 189
      costs, 193
      feedstocks, 189
      manufacturing, 190-192
   fuel economy, 203
   methanol, 194, 195
      chemical composition, 194
      costs, 195
      feedstocks, 195
   Savannah River Plant program, 206-208

B

British Thermal Unit content
   compressed natural gas, 143

British Thermal Unit content *(Contd.)*
   ethanol, 189
   hydrogen, 209
   liquid natural gas, 178
   methanol, 189
   propane, 13

C

Canadian energy requirements
   coal, 9
   electricity, 9
   natural gas, 9
   oil, 8, 9
   propane, 9
   renewables, 9
Canadian energy reserves
   natural gas, 7, 8
   oil, 7, 8
Canadian propane supply, 11
Century propane equipment
   mixers, 48-50
   vaporizers, 46-49
Compressed natural gas
   BTU content, 143
   compressors, 171-174
   emission levels, 147, 148
   filler hose, 155, 157
   filling facility, 152
   fuel cylinders, 155, 156

Compressed natural gas *(Contd.)*
   fuel system, 156-158, 161-163
   Italian experience, 149-153
   meters, 175, 176
   mixers, 156-158, 160, 161
   pressure regulators, 156-160
   regulations, 165-169
   safety precautions, 169-170
   United States experience, 147-149
Computer Command Control system
   computer fix control, 134
   input sensors, 115-120
   management functions, 120-129
   purpose, 115
   self-diagnostic system, 129-132
   service precautions, 132-134
Cryogenic fuel tanks, 178-180, 211-213

D

Diagnosis, propane systems
   detonation, 114
   hard starting, 111
   hesitation on acceleration, 113
   low fuel economy, 113
   misfiring on acceleration, 113
   power loss, 113
   rough idle, 111
Diesel boosting
   alcohol, 202, 203
   propane, 63-65
Distributor advance requirements
   alcohols, 201
   compressed natural gas, 143
   hydrogen, 212
   propane, 100-108
Dual curve ignition controls
   compressed natural gas, 143
   propane, 143

E

Emission levels
   alcohols, 203-206
   compressed natural gas, 147, 148
   hydrogen, 218
Energy cost factors, 7

Energy requirements
   Canada, 9
   United States, 4, 5
   world, 1
Energy reserves
   Canada, 7, 8
   United States, 2-4
   world, 1, 2
Energy sources, 5, 6
Engine tuning
   alcohols, 199-201
   compressed natural gas, 143-147
   hydrogen, 211-217
   propane, 91-110
Ethanol (*see* Alcohols, ethanol)
Excess flow valve, 69

F

Filler valves
   compressed natural gas, 157-162
   liquid natural gas, 179, 180
   propane, 69
Filling facilities
   compressed natural gas, 151, 152
   hydrogen, 214
   liquid natural gas, 185-188
Fuel economy, alcohols, 203
Fuel lines
   compressed natural gas, 167
   hydrogen, 215
   liquid natural gas, 178, 179
   propane, 68
Fuelocks
   electric, 79
   vacuum, 34, 35
Fuel systems
   compressed natural gas, 156-158, 161-163
   hydrogen, 211-220
   liquid natural gas, 179-184
   propane, 72, 76
Fusion energy, 224-227

G

Gann distributor plate, 106
Garretson propane equipment
   mixers, 37-40
   vaporizers, 35, 36
Gasohol, 192

Index

H

Hydrogen
 BTU content, 209
 cost, 210
 exhaust emissions, 218
 filling facilities, 214
 fuel systems, 211-220
 fuel tanks, 210, 211
 safety aspects, 218
 sources, 209, 210
 weight, 209

I

Ignition requirements
 compressed natural gas, 143-146
 propane, 91-99
Impco propane equipment
 fuelock, 34, 35
 mixers, 26-34
 vaporizers, 21-25
Infrared testers, 110

L

Liquid natural gas
 BTU content, 178
 filling connections, 179, 180
 filling facilities, 185, 188
 fuel system, 178-184
 fuel tank, 178, 179
 liquid mode operation, 180-182
 liquefaction process, 177
 vapor mode operation, 182, 183
 wiring diagram, 184

M

Methane (*see* Natural gas)
Methanol (*see* Alcohols, methanol)
Mixers
 compressed natural gas, 156-158, 160, 161
 propane, 26-34, 37-40, 44-45, 48-50, 58-61

Mixer adjustments
 Garretson, 37-39
 Impco, 28-30
Mixer selection, 31-33

N

Natural gas
 air fuel ratio, 146, 147
 chemical composition, 143
 cost, 141
 ignition requirements, 143-146
Natural gas availability
 Canada, 141, 142
 United States, 139
 world, 137, 138, 140

O

Ocean Thermal Energy Conversion, 227

P

Petrosystems propane equipment
 mixer, 61-63
 vaporizer, 64
Pressure regulators
 compressed natural gas, 156-160
Propane chemical qualities, 13, 14
Propane conversions, computer systems
 computer fix control, 134
 precautions, 132-134
Propane conversion costs, 18
Propane conversion equipment
 fuelocks, 34, 35, 79
 mixers, 26-34, 37-40, 44, 45, 46-69, 51-54, 63, 64
Propane diesel boosting, 63-65
Propane engine design requirements, 17, 18
Propane fuel systems
 dual fuel, 76
 straight propane, 72
 wiring diagram, 81
Propane fuel tanks
 excess flow valve, 69
 filler valve, 69

Propane fuel tanks *(Contd.)*
    magnetic fuel gauge, 70
    outlet valve, 69
    relief valve, 70
Propane regulations
    hose requirements, 87, 88
    motor fuel tanks, 83-86
Propane safety precautions, 68
Propane supply
    Canada, 11
    United States, 11, 12
Propane tuning requirements, 14-16, 95-111

R

Relief valve
    compressed natural gas, 158, 161, 162
    hydrogen, 213
    liquid natural gas, 179, 182, 183
    propane fuel line, 71
    propane fuel tank, 70
Remote fill hose, propane, 85

S

Safety precautions
    compressed natural gas, 169, 170
    propane, 88
Solar energy potential, 223

T

Tartarini compressed natural gas equipment
    mixers, 160, 161
    pressure regulators, 159, 160
Tartarini propane equipment
    mixers, 44, 45
    vaporizers, 41-43

Tuning requirements
    alcohols, 199-201
    compressed natural gas, 143-147
    propane, 14-16, 95-111

U

United States energy requirements
    coal, 45
    natural gas, 4, 5
    nuclear, 5
    oil, 4, 5
    other, 5
United States energy reserves
    coal, 1-4
    natural gas, 1-4
    oil, 1-4
United States propane supply, 11, 12

V

Vacuum switches
    liquid natural gas system, 184
    propane system, 80
Vaporizers
    ECl device, 23-25
    ECl device adjustment, 110-112
    propane, 21-25, 35, 36, 40-43, 46-49, 51-54, 63, 64

W

Wind energy, 223, 224
Wiring diagrams
    liquid natural gas system, 184
    propane system, 81
World energy reserves
    coal, 1, 2
    natural gas, 1, 2
    oil, 1, 2